HARBO WORLD

HARBOURS, GIRLS AND A SLUMBERING WORLD

Written and illustrated by
JAMES PAGE-ROBERTS

The Mudlark Press

British Library Cataloguing-in-publication data

A catalogue record for this book is available through the British Library

First published in 1999 by The Mudlark Press
PO Box 13729, London W6 9GN

Cover design by Robert Page-Roberts

Scanning by Rachael Tremlett & Ray Leaning

Design and typeset by Ray Leaning,
Smart Credits, Brighton

Printed and bound by Biddles, Guildford, Surrey

ISBN 0 9530517 4 9

CONTENTS

I dedicate this book to the late Anna de Goguel to whom I wrote the letters, and to her son, Constantine, who returned them to me after her death.

Books also written by the author include:

GUIDE TO A DOCKLAND OF CHANGE
(The Mudlark Press)

CANARY WHARF AND SIGHTS FROM DOCKLANDS LIGHT RAIL
(The Mudlark Press)

DOCKLAND BUILDINGS OLD AND NEW
(The Mudlark Press)

COOKING IN DOCKLANDS PAST AND PRESENT
(The Mudlark Press)

VINES IN YOUR GARDEN
(Argus Books)

THE BEST WINE IN THE SUPER MARKET
(Foulsham)

The first three editions of
THE BEST WINE BUYS IN THE HIGH STREET
(Foulsham)

THE OLDIE COOKBOOK
(Carbery Press)

VINES AND WINES IN A SMALL GARDEN
(The Herbert Press/A&C Black)

WINES FROM A SMALL GARDEN
(Abbeville Press, New York)

DRUIF EN WIJN UIT EIGEN TUIN
(Schuyt, Haarlem)

THE DRAWINGS

Highly successful exhibitions of the author's drawings were held at the Reid Gallery in Cork Street, London, from September 29th to October 15th 1960, and at the Kintetsu Gallery, Osaka, Japan, from October 13th to October 18th 1961.

The illustrations in this book have been selected from those still held by the author and from private collections.

The indicative maps on pages 252-254 have been drawn by the author. They are not to scale or related to any projection.

Written and illustrated without prejudice, this account of a voyage of discovery was conducted during an unusual period of history when the world was at peace.

Printed in ***bold italics*** are extracts from letters sent back to England from abroad, and in roman type are thoughts at the time and matters believed to be inappropriate for letters.

CHAPTER 1

Birkenhead

The great trading ship stood there alone in the grey countryside. There seemed to be no buildings of note nearby and little in the way of dock paraphernalia.

Large ships often look quite out of place when seen rising out of a landscape, more especially out of a flat and bleak Birkenhead one like this. She looked in fine trim, well painted, no rust, clean flags. The Red Ensign hung limply at her stern in the foggy drizzle, as did the Blue Peter aloft to tell all that she was about to leave for muddy estuarine water and clear sea. The flag that tells when a pilot is aboard was not to be seen. But then I was early. When the taxi drew alongside, I confirmed that the name of the ship was 'Achilles'. The first stretch of a journey that I hoped would fill a gap in an education destroyed by war, and at the same time give me a certain self-assurance in both worldly matters and matters of the world, had gone without mishap.

CHAPTER 2

A Short Past

I could hardly complain of having led a cloistered life as an excuse for 'seeing the world'. Left alone to fend for myself in the country during school holidays had taught me not to fear the night, to walk silently, and how to feed myself. Then, as a war refugee in New England, in the USA, I learned both to live among foreigners and, as a volunteer fireman, about knots and how to carry a ladder. My school and trade school education there counted for little. But I was observant. And to be so is an education in itself.

At last old enough to join up in the RAF, I boarded a merchantman in New York, bound for Halifax in Nova Scotia, Canada, there to join a convoy for an Atlantic crossing to Liverpool. Our vessel broke down once or twice off Cape Cod on the leg north up the coast. Once off in the convoy, we proceeded at the pace of the slowest ship, which may well have been ours, though there were quite a few three-island coal burners in our number. We were one of perhaps fifty or more ships, and positioned vulnerably on the outside. A lone destroyer accompanied us outward. In mid Atlantic, I came into the mess room for breakfast to find my few fellow passengers in various states of agitation. We had been attacked by German U-boats in the night, and several ships that had been with us the day before were now absent. Everyone on our vessel had stood by the lifeboats throughout the attack. Sound asleep in my bunk, I had missed it all. It is a huge advantage to be lucky by nature.

The RAF accepted me for pilot training and sent me back to 'civvy street' to wait my turn. Many pilots were being killed at that time, so it was most important to train replacements and also to have more available for future operations. But there were only facilities for training a certain number at a time. I would be contacted in due course. So, to help with vital food production, I took a job as a farm labourer. After a while I heard that prop-swingers were wanted at an airfield some five miles away from my lodgings. I bicycled over and was engaged. Each day I swung propellers to start the engines of Tiger Moth biplanes. 'Switches off, petrol on, throttle closed,' I'd shout, turning the prop to suck fuel into the cylinders, 'Contact!' I called out. 'Contact' would come the reply. Then I'd flick the propeller around, leaning-cum-falling away to avoid being minced as the engine started. Then I'd drag the chocks away from in front of the wheels (the aeroplanes had no brakes) and hold a lower wing tip to help guide 'plane, pilot and pupil on their way over the ground toward where they would turn into wind and take off.

The management came to know that I was waiting for flying training. As we then had no way of knowing what forthcoming weather over the airfield might be, an instructor would fly a Tiger Moth up-wind for some time to discover if any noxious rain or clouds might be approaching to make flying dangerous for the beginners. On these weather flips the second cockpit was usually empty, so I was sometimes taken along. In so doing, I obtained flying instruction. When I did come to fly officially, it was thought that I was a natural in the air, flying solo in what was thought to be a record time.

So, after basic training, first solo flight, and a spell mending the roofs of bombed houses in London's dockland, they sent me back to America (Oklahoma this time) where, after escaping unharmed from a mighty crash, I gained my wings shortly before the end of the war.

After my return to England and a bout of tuberculosis, contracted in cold and damp conditions when food was scarce while waiting to be demobilised, the disease returned after the start of a career in medicine. Complete rest was proscribed, combined with the only known 'cure' at that time (a needle thrust between ribs each week to allow air to enter an induced gap between lung and rib cage). My career in medicine had ended. I then went to art school and on to the Old Vic School to learn about theatrical set and costume design, and still managed to travel abroad, having the chest air topped up in conditions both hygienic and otherwise. Then, after learning to paint scenery at the Covent Garden Opera

House, I designed for repertory and touring shows, television (black and white), ice shows and such. This work was hard, satisfactory, spasmodic, badly paid, and did not lead me to the upper reaches of the business.

I had had the good fortune, or perception, to buy a bombed-out house in the London Borough of Fulham. Now, after seven years, I had reached a stage when it was no longer necessary to visit hospitals for lung treatment. Work in the theatre held no future for me. I just wanted to paint and draw. So why not, I thought, sell up and see the world, painting and drawing as I went?

I sold the Fulham house for a handsome profit, and bought a wreck of a cottage in the country to have a foothold in England. This place would have a studio with a marble floor made from the washstand tops I had bought and saved for such a purpose. Otherwise, ideas for developing it would be gleaned, perhaps, from observations made in other countries.

CHAPTER 3

The Kit and Preparation

To travel light and to travel alone. Those were to be my maxims. The main, and heaviest items of luggage to be taken, were materials for drawing and painting. So, to reduce weight, packets of my favourite quality of paper were dispatched to postes restantes in various parts of the world through which I might pass.

Into a lightweight, expandable, fibre suitcase with a good lock, I put a small kit bag, mosquito net, a mechanical wind-up shaver, a tooth brush, an anti-snakebite kit (crystals and knife), string, blow-up pillow with a waterproof pocket in which to hold valuables, a bottle of chalk and opium (supplied for stomach trouble), a villainous-looking knife for protection and sharpening pencils, a tropical suit, condoms (including a washable, reusable one), anti-malaria tablets, a map of the world, a white hat (which I somehow forgot), sun glasses, oil paints, brushes, paper, ink and pencils. Although rather a lot to cart around, perhaps much could be left in the suitcase at "base camps". The warm clothes on my back I would throw away or send home on reaching hot climes.

With this suitcase in my hand, traveller's cheques, passport and a boat ticket to Tokyo in my pocket, I made my way by rail from London to Liverpool.

CHAPTER 4

England to the Red Sea

23rd December 1958

We sailed at 6 o'clock and have now reached the lock gates (8 o'clock).

My fellow passengers consist of one biochemist, one bishop and wife, and two women from the north of England with four screaming and ill-behaved children.

Dinner was served with me sitting between the two women and their children, with one child spitting ice-cream at me. Although my ticket will take me to Japan, I certainly do not intend to go that far with this crowd. Could cruises be like this?

The biochemist saves the day. He is returning to work in the rubber industry in Malaya, and has lots to talk about. We have agreed that if we cannot stand the children further, he'll teach me biochemistry and I'll teach him painting.

The bishop is a very nice old boy, whose opening gambit was: 'You must be between 25 and 29.' (Actually, I think I'm 33.) Perhaps such underestimating was part of his training for the cloth.

The Captain's skills of seamanship are held to ridicule by his officers. Fortunately the skipper seems to be more interested in deck games. Opinion has it that the other boats of the line, staffed by Chinese, are the best for food. Perhaps we have British Rail trainees as cooks. Wonderful ingredients are woefully treated in the galley.

Being mean with the use of paper at this early stage, I'm limiting myself to a few drawings of related parts of the ship's superstructure. They remind me a little of those done by Paul Nash on his way to America. Jim

A rough and cold crossing of the Bay of Biscay coincided with Christmas Day and an issue of seasick pills. The Bishop held a service and the Captain stood everyone a glass of wine.

I had high hopes for the future success of my newly acquired tropical suit when, during a particularly nasty roll of the ship, a glass of port slid down the table and deposited its contents all over it. I put the jacket and trousers under a hot tap and hung the suit to dry. It recovered beautifully. I would still look fairly presentable should I contact any of the smart people in distant lands to whom I had introductions.

1 January 1959

I saw that several of the crew were drunk last night when the New Year was ushered in. It seems there was hell to pay this morning as we lay at anchor in Port Said. Someone must have been sober enough to navigate the ship. Perhaps it was a pilot alone.

The natives are aboard en masse. I bought a cheap sun hat as a temporary measure, but the goods offered are shoddy in the extreme. Suggestive postcards are sold to us tourists along with those depicting the sphinx and the statue of Ferdinand de Lesseps. 'Filthy' books are offered. I looked at one, and every other word seemed to be either shit or piss, which I suppose is appropriate for a port generally known as the anus of the world. So I took great and appropriate pleasure in using the lavatory this morning.

A sailor fishing from beside a lifeboat has had his bait snatched from right next to him by a swooping kite. Clearly I will have to watch my valuables from now on.

The weather is nice and warm at last, and the strange, dung-coloured houses with balconies, set among palm trees under a blue sky, with Arab dhows and boats of many colours floating in the blue water in front of them, makes a pleasant, picture postcard sight.

We were under way at ten in the morning and are starting down the Suez Canal at a man's running pace. I know this as for some reason a man is actually running alongside the ship.

There is a lot of canal widening going on. The sight of thousands of Arabs carrying sand and soil in buckets must be the same as in biblical times. Sometimes there is a hooded and black-dressed woman to be seen, and occasionally one or two men face Mecca as they kneel on their mats to pray.

Now we are anchored in a cutting with seven other ships, waiting for the next convoy to steam by us in the opposite direction. Then we'll be going on to the next passing spot.

At the next passing place the pilot joined us for dinner and I had the good fortune to be sitting next to him. He told me that he had been a pilot on the canal for two and a half years. Out of the 200 pilots, and now a further 75 recently added (mainly ex British tanker Captains), he was 40th in seniority. Even so, he had his grievances. Having given up his job as a deep sea Captain he had not realised that being at home with his wife so much was a situation not leading to happiness. Also, his wife did not know how much he earned before. Now she did, and made appropriate demands. He was overworked. And although the authorities had promised him leave, he had not been able to take any for more than two months. Moreover, he did not like Port Said. He liked Cairo. But worst of all was that he and his wife were obliged to make regular visits to other pilots and their wives, and even the wives alone if their husbands were at work. Why? Well, it was a British custom handed down, presumably to ensure that no-one became lonely in such an isolated outpost of the Empire.

The desert looks very attractive with its camels, towns and villages, and sun setting over palm-fringed oases. These are my first glimpses of the truly exotic. Sometimes the whiff of human and animal excreta from villages penetrates the vacuum of shipboard life. That, and the wind blowing sand into my eyes, tells me that it's not just a picture postcard world out there, but real Middle East. Jim.

The next stop was to be Port Sudan. Here I would be able to put foot on the African continent. But I was not about to start my travels through Africa. I had also decided that to look at India was too much to include in this trip. It was the Far East that beckoned. And the more I spoke with the chemist and heard about rubber plantations and Malaya, the more I began to think that I would leave the ship in South-East Asia. Looking at my map of the world, with a line already drawn between London and the Suez Canal, the better and more sensible that decision seemed to be.

CHAPTER 5

Port Sudan and the Fuzzie-Wuzzies

JANUARY 1959

We have moved from the Gulf of Suez to the Red Sea, and a warm and stiffish breeze blows from the Saudi Arabian desert.

My chemist companion tells me that he is going to write a book when he returns to Malaya. He has spied his maidservant making love to her boyfriend, and has arranged with her that he would be watching them again — an episode to go into his novel. But what he really likes, he told me, were boys with bottoms like peaches. He has fixed his eye on a sailor, and has become rather moody as he contemplates his ends. On the other hand, one of the women with children is having more luck with another sailor.

Having realised how foolish I was to have entered three deck game competitions, I immediately bowed out of the first (Pin Quoits) by losing to the Bishop. We were both delighted with the result. Jim.

The weather became distinctly hotter as we ploughed through the Red Sea, and everyone was now attired in tropical gear. Many took to sunbathing. A makeshift swimming pool was constructed on deck. I threw away my warm clothes. Our ship was due to dock at Port Sudan around midnight. I was looking forward to it.

With my correspondence seldom dated, because I am usually unaware of such things, I found it to be both a useful and delightful gesture on the part of the Blue Funnel Line that they sent a picture postcard home announcing whenever their ships arrived at or left a port.

4 January 1959

We arrived at Port Sudan early in the morning, but because we are carrying explosives (the first I have heard about it), regulations here state that they must be offloaded into a barge before the ship could be tied up at the dock. So it was not until mid morning that we were able to go ashore.

I obtained sandwiches and, after doing some drawings of Fuzzy-Wuzzies and Sudanese of various tribes getting ready to load cotton aboard our ship, set foot on the African continent for the first time.

There is a ferry to the town proper. I boarded it and, accompanied by 20 Sudanese, a man rowed us across. It took some time.

The first impression one gets of these people is that they smell strongly. The Fuzzy-Wuzzies treat their fuzzy hair with a camel dung and urine mixture, which I suppose accounts mainly for the unpleasant odour. Each has a long three pronged comb stuck into their coiffure with which to scratch. They are as black as your hat, and wear dun-coloured clothes consisting of an Arab style pantaloon with an over smock. This does vary quite a bit, but the colour of their clothes and skin blend so well with the surroundings that when, for instance, two might lie on a tarpaulin, they are almost impossible to see.

There are sects besides the Fuzzy-Wuzzies, and many of them wear turbans. They wear no shoes, and one notices immediately that their toes are used like their hands. They are an extraordinarily graceful crowd - all of them. Even when reclining on the 4 cwt bales of cotton, or just curled up on the ground in fantastic postures, they are extremely elegant. Paid 8/- a day, they seem to be reclining for much of the time. Yet because there are so many of them, the work goes on apace. Sometimes they chant, but it is not an interesting sound.

Now for the town. It spreads out over a vast area with nomadic tribes on its verges. Theirs are interesting encampments, and seem to consist of matting tents in a reed-like compound. Here there are many camels. Buzzards and vultures are to be seen in the sky. It's possible to get quite close to the vultures. In fact, I saw two children chasing away a large one from a drinking trough. They must have been within a couple of yards of it before it flapped its huge wings and flew away.

From the featureless and vastness of desert, herds of goats are being driven toward the Red Sea ferries, and subsequently Arabian cooking pots.

I have seen a small family stop in the desert to make their

ablutions and kneel toward Mecca. It was an impressive little scene of deference to their beliefs. And because they remained fairly still, I was able to draw them.

Trades like to keep together in communities, as do tribes. The Indians have a section. And here I saw a wedding in progress within a long and violently decorated tent.

There is no green except for trees and bushes. Sand is everywhere. Desert and streets are the same, one becoming the other without notice.

A headmaster showed me around one of the schools. He then took me to visit some of his friends in homes where women were heard but not seen. They are seen in public, but are then heavily veiled. Rooms are barren, with tall ceilings. Furniture is sparse. Some houses I entered had electricity and water. But I think these were shown to me to indicate how the Sudanese were moving with the times.

The schoolmaster, who seems to be well known and a man of influence, is very pleased with the present regime. He says that parliamentary democracy will not work in the Sudan. I certainly noticed the state of martial law. There are policemen everywhere. And I was completely searched when leaving the dock. Perhaps that is why explosives are so strictly supervised.

It is their winter now, with temperatures in the 80s and 90s. In summer it is said to be the hottest place on earth. So I seem to have chosen the right time for this all-too-brief visit.

I am now back on the ship. Some English have come aboard to get flour, bread and butter, there being none available here. Jim.

CHAPTER 6

Aden and the Indian Ocean

JANUARY 1959

We are at sea again and a swell is mounting, accompanied by a strong warm wind. Being now fully laden we thrust and plough our way through the waves, and the wind blows spray right over the ship. Flying fish, which I am told are good to eat, are being blown aboard. They are only 5" to 6" long and leap and fly in swarms, looking like little birds or dragonflies. Sometimes they appear to be tying to escape our bows, adding to the dramatic spectacle. Jim.

8TH JANUARY 1959

We arrived in Aden in the early morning to anchor offshore and refuel. Arabs piled aboard with goods to sell, the quality of which was very much better than at Port Said. We were allowed two hours ashore, so I took the chance and was ferried to the land. I wanted a quick look around, and a nice policeman found a taxi for me and even bargained on my behalf for a good price.

Aden has a huge and almost landlocked harbour that was once a volcano. The town proper is built at the foot and up the sides of large, mountainous rocks. It is essentially an oiling-up place for ships that pull up alongside oil buoys in the harbour. They anchor and the oil is pumped aboard.

I went to visit Camp Aden, some five miles from the dockside. The paved road to this sweltering and extinct volcano-cone town had been blasted through the lip of rock. The volcano shape seemed to intensify and funnel the heat inwards. On the way there we travelled through a fantastic mixture of buildings and warrens of hovels.

Whereas the Sudanese women in purdah had their eyes visible, here they are absolutely and completely covered. Their thin veils are sometimes patterned with red and black designs, often giving them the appearance of hideous creatures.

Aden, though mostly an Arab town, has a very strong British stamp about it. The military are much in evidence, and RAF aircraft are constantly taking off and landing from an airfield in the distance.

There was no time to linger and draw, not that there was much to interest me, anyway. And I felt somewhat ill at ease in this very Arab country. Equatorial life beckons.

I have rather overdone the sun on deck, as the north-east monsoon winds kept me from feeling too hot. These winds have preceded rain in the form of gigantic storms. They have burst over the ship by day but have gone by sunset, leaving the air beautifully clear and bright. The combined force of these strong winds and the 16 knot speed of our boat, creates a gale that is difficult to stand up against. The wind and heat tend to dehydrate us, so there has been a ship's issue of salt tablets.

The day's activities have lapsed into routine. Up at 8 o'clock, breakfast at 8.30, change the daily library book, read in a sheltered spot, wash clothes, lunch 12.30, sleep or read until 4, deck games for exercise, shower, change for dinner by putting on a tie and slippers instead of gym shoes, drink iced Bass, dinner, cigar and more drinks, retire, read or write, turn on the air blasts and lock the door two ways to stop it from rattling. All this was to be repeated day after day.

The passengers have become listless and complain of lack of sleep. A pest of a child will not leave me alone. When he pulled off my key chain I gave him a clip. It did the trick, but all hell broke loose with the mother. I have had the gross misfortune to win at a deck game, so have to play again in the competition tomorrow. This is a long stretch, and I can't wait to get on with my travels and drawings. Jim.

Being quite tall, and European, I had already become very aware that on these foreign shores I cut a most conspicuous figure.

The ship plunged on through the Indian Ocean, passing four or five miles off Ceylon, where catamaran fishing boats were at times quite close to our ship. Offshore breezes carried the scent of sandalwood and cinnamon out to sea from palm-fringed beaches, and even into my cabin. These were thoroughly inviting smells, coming from an exotic landscape. They would draw me back at another time. Having reached the coast of the Malay mainland, and then Georgetown, Penang Island, where there was no dock for our ship, we dropped anchor offshore on January the 17th. At last I was about to start travelling at my own speed, in my own time, and in my own direction.

CHAPTER 7

Penang and the First Steps

17TH JANUARY 1959

I think I rather surprised the Bishop with the suddenness of it all when I said goodbye and struggled down to the pilot boat with my suitcase.

With no arranged place to stay, the Captain had suggested that I might try the Mariners' Club. Once ashore, I took a trishaw to investigate the situation, with the cyclist sitting behind the seat. My Indian driver recommended one place as being reasonable and clean. He waited outside with my luggage as I looked inside. The rooms had no doors at all, only the swinging kind as seen in western films. And inside the one room I inspected there was only a bed. This stood on a wooden floor, patterned all over, and especially toward the corners, with short brown marks. It was most decorative, in an abstract kind of way, and I spent a little while admiring it. The hotel was a brothel, and the marks on the floor had been made by the inmates and their clients when throwing away lighted cigarettes before or after encounters. Perhaps they would have even shown me the room, if occupied. Anyhow, whatever went on inside would be clearly visible from the passageway outside. Sex and privacy do not go hand in hand here.

We continued the search, and how splendidly oriental was this cycle trishaw method of transport. Through bustling and colourful Chinese-looking streets, where Indian and Chinese music is played so loudly, I found nothing remotely suitable, let alone either private or secure. But the Mariners' Club, although officially unable to welcome me because I am not a sailor, have given me a clean and secure room with running cold water. I am temporarily an officer in the Merchant Navy.

I have found an excellent map of the town and am about to wander around and take it all in. It seems such a short time since I left cold and foggy England, and here I am in much the same heat as a really hot summer's day in Blighty. I am now in a land so completely different it is almost unbelievable. And it has only taken me 25 days by ship to get here. I already feel very much at home. Jim.

I walked around Georgetown for three hours to get my bearings, lapping up the sounds and smells, and eating Chinese food. Then, having collected my first parcel of drawing paper from the poste restante, and with a good map of the island at hand, I set off by bus to investigate and circumnavigate it. After the first of four changes of bus, I spied a Chinese hotel on the north coast that might do for me later. Then I alighted at a charming little fishing village that unhappily had no hotel.

19TH JANUARY 1959

I think the entire population of a fishing village in the north of the island formed up around me when I was offered lunch by the proprietress of a café. I was hungry and accepted. The meal, taken at a table outside, was made up of two slightly warmed eggs and an empty cup, squares of bread and a fork. With this came two slices of papaya, a spoon and a glass filled with hot Milo — a sort of cocoa, manufactured in England. I must have eaten it all correctly, as I could see no sign of surprise on anyone's face. But I don't like raw eggs.

I thought that the word 'toilet' might be understood, and asked for it. But I was directed past what I thought might be the privy to an area of brick in which there was a pool of live fish. But I noticed a damp corner, so all was well. Perhaps I should have asked for the WC.

The next bus ride zigzagged up and down very steep hills, almost mountains. There appears to be a certain amount of jungle, but nearly every scrap of land that can be seen from the road is densely covered with coconut palms, oil palms or rubber trees. The flat, low-lying pieces are paddy fields.

I am disappointed in rubber trees as I expected them to look like huge rubber plants. They don't. They look more like birch trees and, incidentally, do not grow in orderly rows on level ground, but on any and every bit of soil that will maintain them on the mountainside.

Coconut trees seem to be the most numerous and are gigantic. They are graceful, slender trees and, right now anyway, are furnished with nuts in varying degrees of ripeness. Coconuts don't seem to be picked, but fall when ready. I'd certainly think twice before walking beneath a tree of them.

Other things that immediately hit my eye are that all the poorer houses are on stilts. But both new and old buildings are often lovely, with just the right amount of profile decoration on their fronts. Hula hoops are much in evidence. The population seem to be mostly Tamil Indian and Chinese, with a few Malay, seen mainly in the fishing villages. Most everything is new to my eyes and senses, and it is all fascinating — so exotic.

As if I have not crammed enough into this day, I went in the evening to Miss Peach Blossom's striptease show, where I managed to do yet more drawings. The stage had been set up on some dusty ground, on which stood a few trees silhouetted against a dark sky. Portable lights with coloured gels lit the stage from the front. Stars were bright in a black sky above.

Although sitting tightly crammed between visiting sailors and many of the male inhabitants of the town, I was able to draw unhindered. My neighbours had other matters to occupy their attention.

The show was a long one. There was a Chinese comic act, a singer who sang Chinese and western songs, and several Chinese strippers. In between the stripping were very poor acts put on by the female chorus. I did not understand a word of the long comic act, but the story revolved around an old man obtaining a glass of urine from a pregnant girl and drinking it, with suitable grimaces. Everyone laughed a lot. The singer, a very pretty little Chinese girl, was excellent.

The Chinese strippers came from Hong Kong. Their dancing was ungainly, mainly consisting of suggestive gestures, culminating in revealing only their breasts. The men in the audience were dissatisfied to see that only the upper half was revealed. But, at the conclusion of the final scene, with the entire cast on stage, the lightweight curtain closed and then blew on stage and over the heads of the cast, giving the girls a chance to show all. They grabbed the opportunity, and did it in grand style. A sailor behind me jumped up on his bench and waved his arms, shouting: 'This is it. This is it.' I gather that there is some regulation forbidding total nudity, with a fine of a thousand dollars. Anyhow, the boys left fully satisfied. Jim.

Having tried on my tropical suit soon after arrival in Penang, I felt dressed for arctic conditions. It felt hot and heavy. I took to wearing a sarong as this item of clothing was ideal for the climate. And there was a simplicity about it that was appealing. 'Going native' had nothing to do with it. To wear one was convenient, and made good sense in Penang.

My garment was made of two yards of printed cotton of some 4' in width. This I had sewn to form a tube. To wear one you climbed in, and folded it across in front as if to make a single pleat. When fitting snugly, it was secured by rolling the material outwards from the top. I wore it during the day and night, pulling it up to the shoulders if the night air became cool. Sarongs were worn by most men and women, the latter rolling them from above the breasts.

I investigated the Chinese hotel I had seen and liked from my bus trip to the north of the island. It was a delightful place, resting rather flimsily on wooden piles over sand, just out of the reach of the sea. I took a room that offered reasonable comfort, privacy and security.

A night club singer I had met when repainting scenery for Die Fledermous in York one winter, was singing at a club in Kuala Lumpur. I contacted her and she came to spend a few days with me at the hotel. With a model, tropical sun, shade, sand and sea, not to mention the satay man on the beach who sold wonderful sticks of succulent pork from his little portable grill, it was a most relaxing and pleasant interlude. I have eaten satay many times since, but never has it tasted so good as on that occasion. And to eat it on a tropical beach, amid the scents of wild flowers and

those of the frangipani tree, was exceptional. Other than satay, food could be either dreadful or wonderful. To aim for Chinese food was usually the best bet.

There were men, almost submerged in the sea, pushing shrimp nets. But I was instructed not to enter the water because of the poisonous sea snakes. Perhaps these creatures only attacked westerners.

The days were beautiful and very hot, with the nights becoming quite cool. And everywhere in the houses were charming little lizards called chee-chaas. These little creatures clung to walls or ceiling, were very shy, made a sweet noise, and were considered to be very lucky.

January 1959

I am trying everything to get a feel and taste of the East. I have drunk fresh coconut juice, eaten bananas from trees, drunk sugar cane juice, guzzled satay with its peanut flavoured curry sauce, and tried Har Ko, which are prawns in a thick sauce tasting strongly of prawn and Oxo. I have tasted masses of weird and wonderful tropical fruit that have incomprehensible Chinese names, and suffered no ill effects, other than a few spots, even when eaten from the most unhygienic of stalls. Jim.

I had told a relation in France that I was about to take a trip to the Far East, and been advised to speak to the head of a French rubber company who was on a visit to London. This I had done, and been told to telephone the head of his operations in Malaya if I was in that part of the world. I telephoned and was told that I would be welcome to visit. It was time for wearing shirt and trousers. My sarong would only be worn at night.

At Georgetown Station I bought a rail ticket for Kuala Lumpur. But this was a station with no trains. They started from near to the dock on the mainland across the water. So a boat took passengers across.

I bought a newspaper for the first time since leaving England. Mike Hawthorne had been killed on the Guildford bypass, share prices were lower due to government uncertainty, there was much bad weather in England with flooding in the Thames valley, and we were 'still fighting' in the Test Match.

CHAPTER 8

Kuala Lumpur and a Rubber Plantation

In Kuala Lumpur I found an adequate hotel for the night, and the following day telephoned the boss of the French rubber plantation business where I had an introduction. Despite my call from Georgetown, no-one had heard of me, or that I might be arriving. However, I was told to come over at four in the afternoon to meet the 'big chief', who was at a funeral when I called.

Desmond Newman and I got on famously and, although he had never heard of me, invited me to stay for the night. On the following day a doctor was to be taken in their company aeroplane to one of the plantations some 80 miles to the north. Would I like to go along? Would I! This might not be quite the kind of rough travel I had expected, but what a splendid opportunity it would be to see the workings of a real palm oil and rubber plantation. I could draw the colourful workers.

I moved from the hotel to Desmond's house on top of a hill in Kuala Lumpur and given a room the size of the cottage I had just bought in England.

JANUARY 1959

At the Newman's home my every need is supplied by servants who call me "master". Had I tried a durian fruit? I hadn't even heard of one. Well a durian is about the smelliest of fruits you could imagine. It is quite large and roughly oval in shape, has a very spiky outer covering holding pockets of edible yellow pulp inside. A durian was taken down to the end of the garden for me to try, but as it smelled like a cross between French drains and a decaying body, I couldn't get near it, let alone eat it. Yet many find it ambrosial. There was much laughter at my inability to tackle such a delicacy.

Being a very civilised house, there is a bidet in the bathroom. It has a jet attachment in the bowl. I sat on it and turned on the cold water to find myself almost inflated. Just how the water pressure could be so great on the top of a hill I cannot understand.

My host and hostess had been invited to the Australian High Commissioner's house for cocktails. I was to be taken along, too. Thank heavens for the smart tropical suit. Much pleasant banter was exchanged on the subject of cricket.

After this party, my host invited us all to a Chinese restaurant for a feast of some 10 courses. The red chillies here are delicious, and served cut up in bowls to be picked up with chopsticks and eaten as desired. Only the crisp skin of a sucking pig is eaten. I don't know what happens with the rest. And although beer is available, it is customary to drink whisky throughout the meal. This is very grand living. And who am I to complain? Jim.

The doctor and I were collected early in the morning and taken to the airport. Socfin's aircraft was an Aero Commander, a fine, twin-engine aeroplane. The pilot, knowing that I had flown in the war, allowed me to handle the controls, but the instruments were far in advance of any that I had seen before.

The manager of Lima Blas Estate, Dudley Boyce, was at the airstrip to meet us. For the first day he drove me around the estate to see the rubber and palm oil plantations. The never-ending work

of weeding and ditch digging was being done by gangs of Malay, Tamil and Chinese women under the foremanship of a man. Over the following three days, the women, dressed in brightly printed cotton dresses, and their overseers, made splendid subjects for me to draw.

We had stopped so that Dudley could speak to the foreman of a gang, when there was suddenly an enormous rumpus of women squealing and yelling. I was standing near to our Land Rover when a huge snake came straight toward me at great speed. The next I knew was that I was standing on top of the Land Rover and the snake, passing beneath the vehicle, was making off toward the nearby jungle. Just how I managed to climb up so high and so quickly I do not know.

Still in a slightly nervous condition, we set off again in the Land Rover, when I suddenly felt something like a sharp bite on my upper leg. I jumped. A glowing ash from my host's pipe had blown on to my trousers and burnt through to my skin.

Most white planters aimed to eat curry once a week on Sundays at curry tiffin. This was a grand assortment of curries, fruits, coconut, banana, sauces and much more. But Dudley liked his curry so much that, except for English breakfast, we ate it at all meals, his Indian cook making them beautifully. Every curry was different, and the powder for each was ground by hand between one stone and another.

The weather was glorious by day, with evening storms that cooled the air and soon passed over. Elephants had been very destructive of late, and you cannot keep elephants out of plantations. They are just too large and strong.

The noises of birds and monkeys, big cats and elephants were truly wonderful in the day and at night. Sometimes a mighty clatter would take place aloft as monkeys were on the move. Branches would creak, crack and fall, and leaves scatter. The height of jungle trees I found to be astounding, as was the lack of green undergrowth in the darkness below them. And in this darkness the danger lay not from animals, that had already run away, but from the festoons of strong and viciously thorned creepers.

Although no crocodiles were to be seen in the muddy, jungle-lined rivers, they were there, claiming a native or two every so often when caught washing clothes on the river's bank. The bodies were then taken to underwater caverns by the crocodiles, to putrify and become more succulent.

There were flowering trees everywhere, and orchids in the trees so common that houses were surrounded by them in racks, grown much as we would grow roses in our gardens. Some orchids were scented. Being parasites, all were still connected to their host wood and needed very little attention when kept as houseplants.

Beneath my mosquito net in the darkness, I took great pleasure in hearing insects buzz while I lay safe within. The dressing I gave the net before leaving England was certainly very effective. Dead mosquitoes coated its top by morning.

And what of the other nasty creatures? I was told that there were plenty of deadly snakes about but they tended to run away. Centipedes, however, were very dangerous, and it was wise to look into the bedclothes before getting in, and to check one's shoes in the morning before putting them on.

I saw and drew rubber tappers scraping slanting cuts in the bark of trees to allow fresh latex sap to run down and drip into an ammonia-laced pot.

Orange-coloured fruit grew in bunches in the heads of cultivated palm oil trees. The berries, much loved by hornbills, are attached to a pithy core. Oil is extracted from the fruit by a crushing and heating process. The nuts inside the berries are then sold for the higher grade oil contained in the kernels.

There were many and interesting subjects to draw. I had a wonderful time on the estate. And the house, on a small hill of untouched jungle, was a pleasure to the eye. The jungle everywhere grew so quickly that it was a constant battle to keep it at bay.

I was taken to see a Malay chief, who sat with his saronged wives in a stilted house on a jungle clearing. He, his wives and the villagers, whose houses stood nearby, needed only to grow rice.

Otherwise all their food was obtainable from the jungle, its verges or the river. I was offered, and accepted, fruit and rice, and was warned that I might also be offered a wife, none of whom were particularly comely.

They had constructed a platform high up in the jungle trees. On this, the chief and his wives would spend much time guarding their durians from flying foxes. It really did seem to be an idyllic life, with no-one needing to work at all, except to grow the rice.

I was even flown back to Kuala Lumpur. What hospitality! And more was to come later in a completely unreal episode, when I was invited to stay for the weekend at a fabulous house set high in the jungle at the bend of a river. The huge, thatched building, built in the Balinese style, offered rooms with every modern convenience, including showers and bidets. It was possible to dive into a swimming pool from my own balcony on waking up in the morning. Beyond the pool, and right up to it, lay thick jungle where many a wild creature lurked and hornbills flew. There was wonderful food on offer, the finest French wines (which, for all their class did not taste right in the tropical heat), cigars and all one might want. And for parties, girls were even flown in from Bali. Such lavishness was possible because only a certain amount of the company's profits could be taken out of the country. The rest had to be spent in Malaya. This was one way of spending it. It was all too much for me. I had come to travel and draw in the rough, as it were, not to be treated with such lavishness, however pleasant it might have been. So, thanking my hosts as deeply as my heart could manage, I took a train and bus south to Malacca.

CHAPTER 9

Malacca, Jungle and Villages

FEBRUARY 1959

I am staying at a guesthouse in Malacca. The town is small and nice and, with no train going through, is somewhat of a backwater. Ships of all sizes anchor offshore and are loaded or offloaded by coolies. The barges are antique.

Food here is either curry or Chinese, the curry being hot in spice but not in temperature. All curry spice is ground by hand from fresh ingredients, taking quite some time to accomplish. The result is delicious, and eaten at macken (a meal).

I have found not much of interest to draw hereabouts, except for the making of small pots that are attached to

rubber trees for latex collection. A solid wooden wheel, resting on a spindle at ground level, is spun by a boy potter's feet. On the wheel, a lump of brown clay is deftly turned into a pot in a few seconds.

Whereas the trishaws in Penang had the passengers sitting between two wheels in the front, here they sit in a basket strapped to the side of the bicycle. And although passengers have a canvas hood for protection, I was completely soaked when a rainstorm suddenly struck. I was then much praised for bringing rain to drought-ridden Malacca. We English have our uses. Yet except for Malacca, it seems there is a huge rainfall. The sun shines most of the time. The rain comes in short and heavy showers, mostly in the evening. I am told that the days are getting lighter. Further enquiry provided the information that there is only half an hour's difference in the length of a day throughout the entire year!

But there is a perennial drought - a cultural one. Whereas we might gawp at a strange Malayan animal imported to the London Zoo, here an artist from England is treated in much the same way, as an oddity and source of considerable curiosity.

I have visited Dunlop's Johore Labis Estate. This is very large, as is the hospitality of its owners. But I have done enough drawings of rubber and palm oil subjects and want

to get on to Singapore, where I have been asked by a planter to line up a Chinese girl (with large breasts) for him when he visits in a few week's time.

I continue to be fascinated by the jungle. It is not just the height of the trees and their variety that amaze me. There is the clatter made by monkeys as they move through the canopy, the snakes (I've seen a giant cobra), the vines, many of which flower continuously throughout the year, the brilliantly coloured and/or scented flowers in the trees, and the orchids of many sizes, scents and colours. Nearly all life takes place high up and not low down. Birds of many colours and shapes abound, and call rather than sing. In fact, it is the noise of jungle life that makes it seem even more exotic than it appears - especially at night.

In clearings are the villages (kampongs) where the Malay lives in palm-thatched (atap) wooden homes on stilts. The

inhabitants of these villages do virtually nothing except fish, grow a little rice and sell fruit from jungle trees. There is really no need for them to work. They build their own houses, feed on what happens to be around, wear very little clothing (it is so warm) and have a wife or two, or three, with whom to pass the time. So why work?

The Chinese, on the other hand, do work, and work very hard. That is why they hold the purse strings of the country. However, one race tolerates the other, there being no such thing as segregation. Yet they do not mix. It is a modern, civilised country, a complete mixture of East and West, but with little cultural or architectural past in evidence. Lovely earthenware bowls are given away with food, and some simply decorated bowls for sale are as delightful as any — and cost virtually nothing.

Yet I now look forward to getting to a Thailand full of Thais and a Japan full of Japanese.

In leaving those who run the plantations, and who have been so generous with their hospitality, I feel that, in one sense, I have paid my way, if only modestly. To each I have given a drawing of their choice. Though as books go mouldy or get eaten by insects in this country, I doubt if they will survive. I still have plenty of work from which to select. The best I will send back. The remainder will be destroyed.

I do hope that my paper supply will be replenished from the poste restante in Singapore. Jim.

CHAPTER 10

Singapore and the Chinese Ladies

My thoughts at the time were to go to Japan for cherry blossom time, but most ways there were fully booked. For that reason, I decided to go to Thailand, possibly Cambodia and even Vietnam first. Booking through to Japan from any of these countries might be easier. And they were roughly in line on my map of the world.

Having reached Singapore and found a bed in one of the houses owned by my rubber planter hosts, I looked up the shipping register to find that a coaster, called the Siaoe, would be sailing for Bangkok, in Thailand, in two day's time. Through the Agents I located the Dutch Captain. He was a little the worse for drink in a bar, but agreed to take me in the owner's cabin. We settled on a price. This suited me well, as I would now have a little time to investigate hot and humid Singapore.

17TH FEBRUARY 1959

I met up with my Chinese contacts yesterday. They are a delightful trio. I was shown the "sights" and in the evening ate at some of the many open-air stalls where delicious Chinese food is cooked and served. This is a humming and most colourful place. After eating, I was asked if I might like to see a slice of genuine Singapore Chinese life. Of course I would! I understand that it was a great privilege to have experienced, as Europeans are simply not accepted in this particular aspect of life here.

There is a quarter where Chinese book rooms — 'anything for the use of' - and sort of Geisha girls are transported in elaborately decorated open trishaws at the command of the

occupants who desire entertainment.

Contacting these girls is done by a Madame, and this would seem to be an art in itself, as one cannot see how it is done, or so quickly. The girls in question either entertain in their own rooms or other people's rooms, or are on patrol, or in transit via trishaw.

Any particular girl can be contacted immediately, and can be produced right away or booked in advance according to her popularity. The one my friends booked was heavily engaged, so rather than wait our turn, another was ordered on chance. She was rather a plain girl.

Now let me start from the beginning. We, that is, the three Chinese and myself, cruised around the district so that I could see the transit in operation. Then we alighted at a known venue and climbed narrow stairs to a strong door. Our most knowledgeable companion, a doctor, rang the bell and an iron shutter was raised so that we could be vetted. My presence needed a lot of explaining (fear of police raids) before bolts were drawn and we were let in. Incidentally, both of the other Chinese in our party had tried to gain admittance in the past and been refused. We were led to a large room, decorated in cream, chrome, and acid green, furnished with several mirrors, a bed, dressing table, table and chairs, hat stand and cabinet full of cups and crockery in general. Here the girl and tea were ordered, and we sat down or watched from the balcony for developments. I watched to see how the communication system worked - and was none the wiser. The tea, and then the girl, arrived.

Now the form is this. For $5 you get "short time". This means that the girl entertains with light banter and can be fondled for 15 minutes, the time given depending on her order book. We had $5 worth, which lasted for half an hour or so, as the girl was not heavily engaged. Then, if she takes your fancy, $20 then, or booked for later, secures her 'personal service', or $50 secures her for the night. I got a few quick drawings done while the fondling was in progress. Then we retired.

Later, we ate Chinese food (naturally) and visited a dance hall-cum pick-up joint. Here one could bring one's own partner or pay excessively for a period of dancing time, making one's own arrangements with the girl for returning to her apartment. Business was bad. Anyway, I did not see how it was possible to see enough through the darkness to select with skill.

The main interest in this dance hall lay in the Madame (a Mammy). She toured the tables of unattached males, asking for their requirements, and could, apparently, elucidate on every aspect of her girls' respective skills, deficiencies, arts, crafts and bodily details. Of course, the doctor and I dredged our minds to think up a sexual act that might not be on the menu. Amid much laughter all round, none could be found.

As the customers were mainly European, I made arrangements with the Mammy for my rubber planter friend to meet up with the girl he desired, namely, a Chinese with large breasts. Whether it will be successful, or whether she had any large-breasted girls on her books, I have no idea. But at least I have done my bit.

Then we drove to the East Coast of the island for the restorative of sweet soup - a bowl of hot sweet liquid with various sweet cereals floating or sunk in it. I can't say I managed much. During this late night repast, a lady hawker offered to sell us ointment to "promote the extension of enjoyment for males".

When I told the Chinese doctor, a professor and a man of many degrees, that I would be going to Thailand, he told me to beware of the Thai women. 'They can cast a magic spell on you,' he said. The spell may be in their charm.

A very nice Frenchman, called Monsieur Merveille, has come to stay at the "planters' house", and with his Chinese friend, Mr. Liaw, we have visited the Tiger Balm Gardens, contrived by a deceased millionaire and now open to the public. Built on a hillside overlooking the sea, it consists of enormous, garishly painted concrete figures of people and animals, set among acidly painted concrete rocks. Caverns depicting moral themes abound, all containing horrific figures with the accent on blood and sex. Surmounting all on the hill, are the phallic tombs of his family, who made their fortune selling Tiger Balm. Even many Europeans swear by the efficacy of this pungent camphor ointment as a muscular cure-all.

The evening was concluded with a show of blue films. Well, they were light-blue, with a scene of a local beauty queen showing pubic hair protruding from her swimming costume causing the greatest gasps of astonishment. Food and sex seem to be the themes of hot and steamy Singapore — perhaps of the Far East.

I have changed some money into Thai currency, and because of the strange letters and numerals, cannot make out if a note is worth one or a hundred pounds. As this is an indication that I will have trouble conversing in Thailand, I am making out a list of my basic requirements in life, like "Where is a cheap hotel?" and "Where can I find a cheap place to eat?" and others. Then, whenever I find someone who speaks English and one of the languages I might have to use, I will write down the phrases phonetically. "Cheap" is an important word, as I may be travelling for some time. Jim.

CHAPER 11

Bangkok and the Brothel

Shortly before the scheduled time of departure, I negotiated a price with a boatman to take me far out into Singapore harbour where the Siaoe lay at anchor.

I was shown to the owner's cabin. An owner's cabin sounded good, but I don't think an owner would have been seen dead in such a bleak, barren and stiflingly hot tin box.

However, off we went on the voyage north toward Thailand, sustained by virtually inedible food, but cheap alcohol.

It was time for bed on my first night, though it was too hot for sleep. I lay down, and felt a slight tickling on my legs. On looking down with caution and apprehension, I saw that an ant run passed over my bunk — and right over my legs. They were large ants, going about their business on their way back and forth between a nest in the hold below and food in the galley next door. There was only one solution. I pulled a sheet over my legs where the ant run crossed my bunk, and we all lived together in harmony for the remainder of an uncomfortable trip.

27TH FEBRUARY 1959

We have just anchored in the muddy river close to Bangkok. And no sooner had the ship been secured to the bottom, than boatloads of girls and their Mammies tied up alongside. After satisfactory negotiations had taken place between the sailors and the Mammies, the girls scampered up the ladders like a jolly crowd of monkeys.

Sanitary arrangements for those on these visiting brothel boats could not have been simpler. A wooden box with a hole in the bottom was placed outboard and secured to the gunwale. It was as simple as that.

Once ashore, I found a taxi driver who could master a few words of English. I told him of my requirements. We visited several hotels, none satisfactory, until we came to a charming one that was half boat on the river and the rest a land-based courtyard of tables, dance floor and trees, with a surrounding balcony which led to rooms, or rather cubicles. A good-looking Thai woman, who spoke perfect English, showed me one of these rooms and I took it right away. There was something immediately appealing about the place. I am so pleased to have found it, and have settled in immediately. That it is a well-run brothel makes it all the more interesting.

I have found a good map of the city and have walked and walked, despite the enormous heat. There are certain aspects that stand out. Saffron-robed monks with shaven heads are everywhere to be seen, many smoking. Some look as young as ten. And the noise of car horns is deafening. Some drivers just keep their horns on

all the time. There is so much noise that no-one seems to take any notice of it. Crossing a road is dangerous, and elicits the full blast. Taxis are mainly little Austins — just right for the size of the people.

There is a lot of traffic on the river with the owners of boats feeding the floating market or just moored to the shore selling wonderful-looking vegetables, fruit and other goods, such as elaborately patterned silver, crocodile skins turned into all sorts, and Buddhist charms of many descriptions. And although many canals intersect the city, few are navigable. These are really sewers, and smell like it.

Dogs make a sorry sight. They seem to be riddled with disease, and lie about at street corners with hardly the strength to stand.

The famous temples are somewhat of a disappointment when seen close to. However, they are splendid when seen from a distance — or on a postcard. There's a great reclining, golden Buddha, taking up the entire inside of a temple, a temple faced with broken crockery, and best, the temple complex of the Emerald Buddha. Encompassed by a colonnade, and by walls decorated with interesting murals, the Buddha himself is set so high that he is impossible to see. Emerald he may not be, but an unknown material of emerald type. The saffron-robed priests are more than friendly, and want to try their English on me at every occasion. I avoided the one who invited me to his temple, saying how tall and handsome I was.

The people have huge appetites. I accompanied one of the brothel girls, past where prawn crackers were being put into bags, to a café near to our establishment (the hotel floating restaurant is expensive) where the girls and I sometimes eat. Very small in stature, she wore a skin-tight black dress. I stood her lunch. She ate two huge plates of rice, two bowls of curry, one salted egg, one hard boiled egg, and topped it off with two glasses of iced tea. There must have been natural room inside her for a quarter of this volume, so I actually watched her swell, almost like an instant pregnancy. It was quite amazing.

Eating habits are quite different from ours. The face is positioned over the plate or bowl and a spoon shovels in the food. What does not get sucked in, falls back on to the plate. The food is either hot-and-spicy, or very sweet. The sweetness is extreme. There is a dish that could be made of some cooked rice, over which has been poured 1/2 pound of sugar dissolved in 1/2 pint of water.

The eating places have their food cooked and on display. This is lovely for me because I can just point. The tepid food is then put on a plate and served with rice and a free glass of iced tea. Rice is the staple background to all main courses. Chicken curry is the most popular dish and is made with a chicken that has been axed through and through until the pieces are very small. It takes some skill with tongue and teeth to separate the meat from splintery bones. The bone bits are placed on the table in smarter establishments, or spat on to the floor in others. There is one main snag about eating these delicious curries. Sometimes, hidden away in the dish, is the most vicious of green chillies. Though very small, it is renowned for being the world's hottest. And if you do bite into one there is no cure. Nothing can assuage the burning sensation — water, tea, beer — nothing works. You just have to wait — in agony, and in my case, amid much laughter from the locals. Prawn crackers are eaten in quantity. Spoons and forks as well as chopsticks are used.

Seeing a bottle of something that looked like methylated spirit on my table, I made enquiries as to what it might be.

It looked most exotic. It was fresh water. I have been drinking the iced water placed regularly in my bedroom and have consumed much iced tea. So with any luck I am building up immunity to whatever bugs lurk within it. I think I'll need this immunity as I am going north soon and will dehydrate if I cannot drink whatever is available.

The Mammy here and I have become great friends. She has taken me by boat along the canals and to the colourful floating market. And she is a fund of information, and articulate with it, having been educated in a Penang convent. She has given me the Thai phonetic for my list of essential desires. So I am now able to ask for the lavatory, a cheap hotel, place to eat, etc. To have this basic form of communication at hand is comforting.

I had intended to start by going north by boat on the river. But the water is too low at this time of year for them to operate upstream. And also I am told that I would be sure to be robbed if I travelled that way. So I will set off by bus from Bangkok, and the Mammy will come to see that I get on to the right one. I will leave weight and valuables here — in the brothel. Jim.

Brothel life in Bangkok was absolutely fascinating. Firstly I had never visited a brothel before, let alone lived in one, and secondly, I had never sampled the services of professional girls. And in this case the selection was enormous, being some 50 girls, though they were not all working at the same time. Through my friend, the Mammy, I was to learn much about the workings of such an establishment.

My room was just a little larger than a cubicle with a basin. This basin emptied into an open drain that was also shared by the girls' rooms up the drainage slope from me. One substance that passed through my room in the drain with some regularity was quite a lot of a thick, white, foamy, sticky substance. I could not imagine what it could have been. The drain just smelled of drains.

The Mammy looked after the bar. She was not only the bar woman but was in charge of the girls, who could come and go and work as they pleased. Soon after I had arrived, one of them decided to go north to Vientianne, in Laos, because that was the place where corrupt businessmen fled, and therefore was where the money lay.

Though there were Thai girls, Chinese girls were the speciality of the house. The hotel advertisements stated that they offered 'Young Chinese girls from Hong Kong'. The girls were comely but not especially so. Most were of the opinion that women were prettier in the north. I discovered later that this was only because they were considered to have paler complexions in the north. My opinion was the reverse. The darker girls of the south were the more lustrous.

The Mammy was the saleswoman for her girls, and was a very good and amusing promoter. The clients were mostly American servicemen. But she did not know much about the actual performances or details of her charges. In that respect I volunteered to help. And as this was my first experience of paying for sex, it was to be of considerable interest to both of us. I think she gave me a cut rate.

It so happened that I once heard a man talking about the conquest of a woman and describing it as rather like waving a flag around in St. Paul's. No-one queried the size of his flag. But although I knew quite a lot about the sizes of various St. Pauls',

trying one after another made it even plainer that size does matter — and not the male size. St. Paul's in some instances was a cramped space with friction, in others one was inclined to rattle around a bit. But although all was tremendous fun and anticipation exciting, the aftermath, usually so pleasing and cosy when both man and woman relax, just did not exist in this business. Orgasm, get dressed like lightning, grab the money and off. That was the order of things. And very sudden it all was. Paid-for sex like this was really a form of masturbation, and not very satisfactory. Another unsatisfactory aspect was seldom to have a shared language.

There was one sexual aspect that I could confirm, though I had already done so fleetingly at Miss Peach Blossom's striptease in Penang. Old sailors' tales had it that the sexual organs of oriental girls ran east-west, rather than the north-south of their western sisters. I suppose it all started, when tales were told in tavern or alehouse, to make matters from distant lands seem more exotic than they really were. All run north-south. I can confirm it.

The pricing system in our establishment was most interesting. Clients (except for me, a resident) were obliged to secure a dancing partner for each hour of the 8-12 evening session. This cost 50tcs an hour (just under £1). Should a customer tire of his partner after say, five minutes, he could pick another, but would have to pay a further 50tcs for the remaining 55 minutes of that hour. The girls could be bought out at the 50tcs rate per hour not used up in dancing. Now the extraordinary thing about pricing was that after the 200tcs obligatory cost of dancing, the rates for enjoying the girls' sexual services was only 100tcs to 200tcs maximum — and that could consist of an entire night of dalliance.

On one morning (everyone, without exception, went to sleep through the hot afternoon), the Mammy held a union meeting. She told me that it was for the older hands to instruct the newer girls on how they might improve their techniques. One item on the agenda was that a serviceman client of a new girl had 'gone on all night'. This was taken very seriously as being a major lack of expertise on the girl's part. So she was advised on how situations like this should be handled. After all, time was money.

It was inevitable, I suppose, that as the girls knew the Mammy and I were so friendly, I was probably comparing one with another, and that this information was being passed on by me to their boss. So my presence was no longer acceptable. It was time for me to move on. I had spent a thoroughly enjoyable stay at the 'hotel'.

So nice and trustworthy were both the Mammy and the management of the hotel, that I felt it safe enough to leave some money and heavier baggage with them until I returned from the north.

CHAPTER 12

North by Bus to Nakorn Sawan

MARCH 1959

The Mammy at my hotel very kindly got up at 5 am to make sure that I boarded the correct bus for the first stage of my journey north. I was glad that she did, because there was no bus station, no visible bus stop and no sign on the bus that I could read.

The gaudily-decorated bus looked rather like one that might have preceded an American circus troupe at the start of the century. It was freezing cold and the bus was open-sided. I was squeezed into a space on a wooden bench with much friendly help. It was a jolly, if somewhat chilly, start.

The vehicle emitted inauspicious sounds as soon as we set off. The engine stopped. In fact, the driver had to stop and then poke around in the engine compartment several times before all was well. The bus was going flat out by the time the sun rose.

I was jam packed in with soldiers, women, children (at the breast and older), old people, priests, animals and all sorts.

As we left Bangkok I was surprised at the reasonable road surface. After about half an hour we hit such a rough and dusty section that it must surely have been under major repair. But the driver did not reduce speed at all. And I thought that the old bus might really fall apart. The bad patch of road lasted for the following 11 1/2 hours! The surface was so bad that for much of the time I had to rise from my seat, bend my knees, and hold on to the bench in front to prevent my spine from being jarred to pieces. And the dust entering from the open sides and large cracks in the wooden floorboards was so bad that we held rags of some sort over our mouths to filter out dirt.

I was an object of curiosity for the entire journey. This was because not only was I a European, but tall and fair. When a change of bus was necessary (where I could have been directed anywhere, so little did I understand what was said to me), interest in me became so great that I really do not believe a European had ever been seen before in these parts. Western people, it seems, do not travel long distances by bus.

I ate food and fruit whenever necessary. This was brought to the side of the bus for sale whenever we stopped. Everyone wanted to see how I would eat it. Pieces of orange-coloured chicken were popular, and cooked rice came in delightful earthenware bowls that were just thrown away to the side of the road when empty.

There were times when I did cringe and withdraw, however keen I was to eat and drink whatever was available to accustom my body to native germs. Such an instance occurred on this bus trip when a friendly man, who was coughing and spitting blood in a very tubercular way, spattered saliva over my food as I was eating it. I stopped immediately, even though I was still hungry. Yet the same man made great play of washing and cleaning the drinking straw that he was about to use, before going on to eat a piece of cake sold to him by a vendor who was handling his wares with fingers that were also picking at a suppurating wound.

The beauty of this race is astounding. In a road gang, among petrol station attendants or those working in the fields, stunningly beautiful women are to be seen. I am trying to work out just why they are so lovely to look at. Their heads, surrounded by masses of blue-black or brown-black wavy hair, are set on long necks. The skin colour is glowing yellow-brown, the mouths large and sensual, the nostrils flared from small noses and the eyes oval and very dark. Jim.

When the 12 hour bus trip was over, I ended up exactly where I had intended. Well, I thought I had, except that my English-printed map was so inaccurate that all who looked at it hooted with laughter. That I had reached the confluence of the mountain rivers Menam Ping and Mae Yom I could confirm. And as I had not crossed a river to Nakorn Sawan, then I must have been at Paknampo where there was a railway station. Whatever the place, it looked very charming and most interesting. So it did not really matter where I was.

There was a small ferryboat for crossing the river to the town, which I later confirmed was, indeed, Nakorn Sawan.

The river was low, as expected, so I had to climb down a steep bank and walk out to the ferry, with my luggage, along a very narrow plank to where the water was deep enough to support the boat. Then, with other passengers, I was ferried (poled) across to the opium centre of Thailand, if not the world.

Looking at the depth of water in the Menam Chao Phyra River that flowed from there to Bangkok, confirmed that it would not have been navigable by a boat of any size. So I could not have reached Nakorn Sawan by riverboat.

CHAPTER 13

Nakorn Sawan and Opium

MARCH 1959

The river, and the happenings around it, form this small town's visible centre of activity. It is a fine sight. The banks, at this time of year stand some 30-40 feet high. These lead down to a clean bed of sand and gravel. There are also crescent-shaped sandbanks, spread around where fast floodwater has formed them. The banks and water are occupied by countless junk craft, which are either beached, or tied to poles in shallow or deep water. Even the smallest craft serves as a family home. And on the sand all around are many children shouting and laughing as they fly brightly coloured kites. The atmosphere is more like that of the seaside than inland central Thailand.

I have taken a room in the hotel that overlooks all these lovely happenings and the confluence of the rivers. Doors are locked with padlocks. There is a very small room for washing, showering, and use of lavatory. The latter is a china trough with a hooded end and a drain hole. A bowl of water is used for both flushing and showering. And the lavatory paper, attached low down, is, accordingly, wet.

The mattress on my bed is stuffed with kapok and is only about an inch and a half thick. So it is hard. But after that bus trip it feels more like a deep, feather-filled one.

I am in the very centre of the opium business, with pure opium coming from the hills by boat from the two mountain rivers to be marketed and distributed. And although it is still legal to smoke opium here, that will, I am told, soon come to an end. So I made it known to the room boy that I would like to visit a 'den' tomorrow.

With the room boy as my guide I have visited the den. It is positioned just behind a Chinese café, where the cook produces wonderful duck and curry.

It was dark inside the barn-like building, but light enough for me to discern the rows of wooden platforms divided by openwork, wooden screens. There were alleyways in between the rows. The men smokers (I only saw one woman smoking) reclined in pairs, each in rather contorted body postures. When not smoking, they were sleeping, gossiping or reading a newspaper. The atmosphere would have been more like a sedate London club if it had not been for the children running around.

The men were of the older generation, light of frame and flesh, but very cheery and showing no sign of depression.

I was allowed to draw, so gave the money for a pipe's worth to an amiable gentleman who was reading a newspaper.

He agreed to be my model. A crowd of fellow addicts and interested bystanders gathered around to watch my progress.

After a drawing or two, more money was demanded, which I gave gladly. But this cash was for a pipe-full for me. I declined it, more for reasons of hygiene than anything else, as pipes seemed to be shared by all and looked rather murky. But they insisted that I smoked. And such a happy and pleasant crowd they were, and knowing that it took more than one pipe full to become addicted, I accepted. Then the fun began. The reclining racks had been designed for a small race of people, and getting my tall frame into one of them caused tremendous mirth. In fact, once in, I took up an even more contorted shape than the inmates.

My addict friend loaded the pipe with a tiny blob of opium. He put it into an equally small bowl on the pipe, held it over a flame and, to many cheers, I took a few puffs, holding my hand over the mouthpiece and drawing smoke from a hole made between my thumb and curled forefinger. I did not inhale, not being a smoker anyway. The tarry-tasting smoke reminded me of the smell experienced when opening up a new rick of hay.

I have now been back to the den again, this time to draw and to take more note of the ritual. The following is the way that opium is smoked by customers:

Although it is possible for one man to accomplish the preliminaries, two are clearly needed. The opium (pure in this case as it had not yet reached the adulterated, commercial, or expensive state as that used in Bangkok and beyond), a brown, sticky substance, is taken from a phial and a small quantity placed on a stick. This is then heated over a permanent flame (taking the form of a small Victorian oil lamp) and rolled into a small ball on the side of a stiffish palm leaf. After a few more heatings and rollings, it becomes the correct size to be put into the pipe.

The pipe is a tube of about 2" diameter and some 2'6" long. The bowl, with the size and looks of a large machine nut, is positioned about 3/4 of the way down the pipe's length. Some silver ornamentation adorns the area around the juncture of bowl and stem.

In the bowl is a small hole, and in this the ball of opium is deposited from the end of the stick. The stick is then withdrawn, leaving the opium mass deep in the bowl. The pipe is then turned upside-down, so that the flame from the ever-burning lamp can ignite it. Then, through a small hole in the flat end of the pipe, the smoke is inhaled in several deep inhalations, before an exhalation.

The smoking part is over quickly. Then the stem is cleaned with a straw-like stick. The smoker, with his head now resting on a rectangular wooden stand with a slope on its upper surface, reclines to take up a contorted position on the boards of the shelf.

A pipe full cost 7 Baht, which, at 58 to the Pound, is hardly excessive. And the opium is at its purest.

With wives and children wandering around as though they are at home, there is a pleasant feeling about the place and its operations. To call it a den would be a misnomer. Only the twilight darkness of the place, with light being obtained only from all the little lamps, gives possibly a den-like atmosphere. You may wonder how I managed to see enough to draw. Well, they kindly turned on an electric light bulb for me.

I was politely warned by words and gestures, that should I smoke opium 'down there' (in Bangkok), it would destroy my health and virility. I have no intention of doing so. Jim.

There was a curious aspect about the shops in Nakorn Sawan. It was not the cleanest or most hygienic place in the world. In fact, you could say that those things were almost completely lacking. Yet in the shops were arrays of soaps, disinfectants, cleaning and washing materials that far exceeded those available in any large town in England.

Wherever I went, a retinue of happy children followed.

My attitude to the sun was quite different to that at home. Here it was to be avoided for fear that I might be burnt to a crisp. The morning heat was possible, the afternoons unbearable and the evenings pleasantly hot. To wear a hat and to have pale skin made sense.

Beer was expensive at 4/- a bottle, but an excellent meal of Thai curry and rice, with pieces of fried chicken, could be had for 5/-. I aimed to drink tea on the assumption that the water had been boiled.

I took my anti-malaria tablets regularly and put up my mosquito net whenever possible. The largest mosquitoes so far were in Kuala Lumpur, and the most numerous in Bangkok.

MARCH 1959

I thought that my room maid's facial features would make a good drawing. So I approached my room boy/guide with signs and gestures that I hoped would indicate my wishes. He seemed to understand and we set off from the hotel. He took me to a large wooden building on stilts just outside the town, to draw 'womans'. This was the town brothel.

In the building were passages and tiny cubicle rooms, all floored in wooden boards with large gaps between them — presumably for ventilation. The exterior of the building was of wooden overlapping matchboard, nailed to vertical supports. The roof was made of corrugated iron and matting.

With absolute lack of spoken language, my room boy and I did manage to settle on a price for an hour's drawing of one of the girls.

Her cubicle had an unglazed window that could be shuttered for privacy. Stuck all over the matchboard walls were pictures from newspapers and coloured reproductions of singers, such as Elvis Presley. Nailed to the vertical wooden supports were colourful Japanese calendars. A square of linoleum covered part of the floor. This had been thoroughly pierced by high-heeled shoes and legs of the one chair where it covered cracks in the floorboards. There was a glass-fronted cabinet holding glasses and crockery, a few clothes hanging on nails, and a modern radio. The small bed took up most of the floorspace.

The bed was covered with a beautifully patterned sheet, the kind I had often seen in Thailand. Its decoration was of broad, pale blue stripes, running down its length, with a grand pattern of coloured flowers in the middle. There was also a blanket, and a hard, frilly-ended bolster.

My model locked the door with a latch so that we would not be disturbed, but several girls tried to gain entry as I drew. Finally, two of them were allowed in, and I drew them all as they sat on the bed. The new girls wanted to be paid as much as my model, but were quite happy when given a proportional amount. To have spent the time in sexual intercourse would have been far, far cheaper. Such is the lot of an artist! Jim.

I had been trying, unsuccessfully, to obtain a good map of Thailand ever since arriving in Bangkok, but to no avail. And as it was time to be moving on north, I started to make plans and arrangements. I would like to have crossed the river by ferry once more to make travel enquiries at Paknampo railway station, but with lack of the language, and inability to even read the printed word, this would have been a futile exercise. So, with the help of my inaccurate map, and judging distances on it by that already known from Bangkok to Nakorn Sawan, I settled upon aiming for a town on the line called Pisanroe, or Pisanuloke.

Someone at the hotel told me that I should be at the station at 6 in the morning to get the 7 o'clock train. I would then reach my destination sometime in the afternoon. I then asked my room boy to write Pisanroe and other potential destinations on cards. On them I wrote their names in English so as not to become confused. These I handed over to whoever sold tickets for buses and at stations, and they, in turn would invariably turn them over and write the fare on the reverse. It was an idea that worked wonderfully. Travelling became easier now with my destination cards and phonetic translation list.

It was time to relax for my last evening in delightful Nakorn Sawan. In the setting sunlight, the children flew and fought with their colourful kites from the sandbanks, shouting with delight as they did so. Their bamboo and paper confections rose into the sky with the greatest ease, where they were made to swoop and dive and glide in combat. In this cluttered sky flew another kind of kite. These scavenging birds of great wingspan, also swooped and dived, missing the many kite strings as they circled over land and water in pursuit of any morsel of food left discarded or unguarded. They were doing the community a great service, and were completely ignored by the people, even when flying at touching distance.

As I watched all this from the hotel balcony, an incident occurred to astound and delight me. A small band of brigands suddenly appeared on the jolly scene. They could not have been more theatrically dressed than they were. They wore slouch, wide-brimmed, Cavalier-style hats, adorned with coloured feathers. They carried old muzzle-loading guns, long knives and bags that may have carried their worldly goods, or perhaps someone else's. They could have come straight out of a village pantomime or a very second rate film. Having crossed the sand with hardly an eye or an eyebrow raised by those around them, they boarded their junk, and remained there out of sight — perhaps to divide their spoils.

I discovered that several of my fellow guests in the hotel were about to take the same train as myself. So, in the early morning, we walked the plank, balancing on it with our luggage as best we could, to board our little ferry. But the boatman had some difficulty in locating deep enough channels for our rather overloaded vessel. So two women and a boy were summoned from the shore to wade beside us, and to push or pull when necessary as we were poled across.

The train was in, steam was raised, the whistle blew, and I was back on course for more travels of drawing and discovery.

CHAPTER 14

North by Rail to Pisanroe

MARCH 1959

At the Paknampo railway ticket office there were three classes on the train from which to choose. First class consisted of only one seat, as far as I could see. Second class had sprung seats, available only if you were early. Third class offered benches, and was populated by the masses and their animals. I opted for a seat in second class, and got one by the window for a good view of the countryside and its happenings.

The train north, through Thailand's great rice growing plains, stopped at every station to offload goods, so it has been slow going, but very interesting. The train left at 7am and I am here in Pisanroe, or Pisanloke, or Pitsanuloke, at one o'clock in the afternoon. Of the exact date I have no idea at all. Even if I saw it written somewhere, I couldn't read it.

Although everything is now very dry, dusty and hot in this monotonous landscape, in the wet season it must be lush and green with crops. The vast, flat panorama, is punctuated by a few sentinel trees. The villages, and even the roads, are built on stilts. It looks so strange to see boats resting on dry land beneath every stilted house and cattle shed. The houses have walls of matting, and roofs of straw or corrugated iron. Water supply seems to come from deep excavations, though in one village every house had its own cantilever water-hoist.

There is still a little water in ponds, and these seem to be teeming with fish. They are probably ponds for fish breeding and food. I gather that the fish in them look like catfish, and were originally imported from China. With few bones and delicious to eat (I had some in a curry in Bangkok), they can stand extreme drought conditions. At one village, every house had its own charcoal-making oven. They were the only ones I have seen.

Despite the dryness and heat, I have seen an enormous quantity of bird life. Among the varieties I could recognise were buzzards, cranes, herons, duck, bee-eaters, snipe, woodcock, kingfishers, and the lowly sparrow. They had perches in the mango and green-podded kapok trees that line the railway track.

Whenever the train came to a halt, food vendors appeared as from nowhere. On offer were fruits of many varieties (some peeled and covered with flies), rice of several kinds wrapped in banana leaves, candied bamboo shoots, dried meat, and chicken cooked in a yellow coating. The drinks, all served in opened condensed milk cans with the attached lid as a handle, were water, iced tea, or coffee. For next to nothing I could buy a plate of delicious curry and loquats. There was sugar cane, cut as a faceted globe on a bamboo stick, and coconuts. The latter's tops were cut across so that the unpolluted water inside could be sucked out with a straw. Then the nut was opened with a cleaver so that the flesh could be eaten.

I write this in a simple hotel, which is, like all other hotels it seems, a brothel. So the rooms are occupied mostly by ladies and their clients. One of the clients, entering the room opposite mine, was a most ancient man, which may say something about the diet, the climate, or the desirability of Thai women.

One room in the hotel has been taken by a prostitute from Bangkok who has come here to visit her small child — presumably cared for by grandparents. She allowed me to draw her as she undressed and changed into other clothes, just as if I was not there at all. I suppose that dressing and undressing in front of men was part of life to her. She kindly instructed me in the use of the ubiquitous and vital bowl. I will buy a nice one as soon as possible, and carry it with me at all times for food, drink, showering and flushing the lavatory troughs. Jim.

CHAPTER 15

A Wood-Burning Train to Lampung

MARCH 1959

I left instruction to be woken at 6 o'clock in the morning so that I could catch the train. But I was woken at 12.30 at night. Then I was woken at six, and had enough time at the station to look at the magnificent, wood-burning, Japanese-built engine. It was kept in sparkling condition and, as I was to learn later, consumed a prodigious amount of wood in negotiating the mountain country. At every station, the tender was filled up from a pile of logs by the side of the railway.

A Buddhist monk had picked the seat opposite mine. He was clearly a man of some importance as he was bade farewell by two very dignified gentlemen who were, I gather, Ministers of something or other.

As we left Pisanroe, I noticed some hardcore being put on to a road surface. As only a tank would have been able to negotiate it, I was now very glad to be travelling north aboard a train and not a bus.

It seems to amaze people that I can travel so light, with only a small, half-filled kit bag. And that I travel with absolutely no knowledge of the language is also a cause for wonderment. But my phonetic list of phrases has already been useful, and I have been able to draw pictures for most of my wants.

On reaching the first hilly section of the track, some wonderful characters came aboard, wearing black, fez-like hats that sprouted tufts of scarlet wool. Each wore a large, solid silver necklet, and the men were armed with antique swords in sheaths.

I notice that the people seem to vary a tremendous amount in their looks, with the exception of old women, who all have their hair cut to an Eton crop and look like rather wicked schoolboys. And it seems that in every province they think their local girls are the prettiest. There are certainly some very attractive ones around.

The monk and I soon engaged in conversation and, as he spoke excellent English, it was a pleasure to converse, after being unable to do so for so long. He invited me to stay at a monastery in a place on the line called Lampung. I was quick to accept. Jim.

CHAPTER 16

Lampung and the Monastery

MARCH 1959

When we arrived at Lampung, the saffron-robed monk and I set off for the monastery in a trishaw. I have learned that this holy man pays for nothing at all, it being a privilege to provide him with all his needs. With everyone bowing and scraping I feel a little ill at ease when I am with him, especially as I do not believe in an after life — either his or mine.

The monastery consists of a bare earth compound, two temples, houses and rooms for novices, rooms for monks, and eating rooms. Then there is a bathroom, a lavatory working on the usual system, a religious altar, and the Abbots room with its reception room. In the compound are more lavatories, a modest flower garden with roses in pots, orchids in pots, a well with a cantilever water hoist, and a trough in which to bath.

Since monks do not eat after mid-day until the following morning, a supporter was summoned to entertain and feed me. This supporter and his cousin, with due acknowledgement to their great kindness, were particularly bland people, despite their glowing robes.

I was offered a bed, or rather bed space, in the temple. Then, fortunately, this was changed to the Abbot's reception room. Up went my mosquito net, secured with difficulty by long extensions of string to anchoring points. This operation was of much interest to the novices. I do not like sleeping directly on the hard floor. But my blow-up pillow, containing passport and money, is a boon at night, and anything soft is placed around my hip area.

All the objects in my bag are regarded with great interest and much giggling. But the one item that caused the greatest astonishment was my not-very-successful wind-up shaver. I shave daily, but the occupants of the monastery have their beards and heads shaved once a month. My monk, having a fast-growing beard, has his head and whiskers shaved once a week.

I find it disturbing that this top monk, being one of considerable power and influence, is at the same time a beggar. The monks' every want is catered for by their supporters. Mine even gets his supporters to run his taxi service, all for free, of course. He keeps saying how much he enjoys life. Here everyone is isolated from life. They are gentle, good, and appear almost homosexual in relationships with one another. No-one seems to have anything to do. I find it a bit creepy, and feel that I am living in a vacuum. Having come to draw people and people in landscape, and done so here, I want to escape — and quickly.

It was most fortunate and convenient that my monk was going to Chiangmai the following day. So I was quick to accept an offer to travel with him.

I was allowed to share the monk's 'supported' breakfast. As one does not share the same dish as a holy man, my food was put into a separate container. There was cooked egg, looking like a tomato omelette, cut into small pieces and cold. There was soup, consisting of beans and mince. But the interesting part was the rice. It was glutinous rice, and easily moulded into a ball in the hands, and then used as a sort of spoon. Morsels of spicy food are pressed into a concavity of the ball with fingers, and then popped into the mouth. The rice is soaked overnight and then steamed. I really need a bowl of my own.

Before leaving in the morning, I was asked to give a talk about England to schoolchildren. I made up a story about a blackbird and this was translated as I spoke. Then I gave a demonstration on how to draw. Finally I was asked to sing. So it was most appropriate that I was able to render the blackbird song, sung in the Berkshire vernacular — 'boogers' and all. There was much applause — led by the schoolmaster, though. Then I was free. Jim.

CHAPTER 17

Through Smouldering Forests to Chiangmai

The train was crowded when we arrived at the station. My monk held back as I forged ahead to lay claim to a second class seat. But he stayed, stock still on the platform, even though he had a ticket. Then I realised that he did not want to push his way through the crowd. So I offered him part of my seat and ushered him aboard. Then someone decided not to go on the train after all, so we both had seats.

The journey up and down the arid mountains was very hot all the time, even though we were sometimes at an altitude where it should have been cooler. Hot whirlwinds sucked ash and other debris into the sky from around the feet of teak trees clinging to burnt slopes. In the dried riverbeds were trimmed boles of hardwood waiting for the rains to wash them downstream toward industry. Smouldering forest fires filled the train with a sickly, incense-laden scent. There were elephants at work dragging tree trunks, walking on the hot cinders of the forests as they did so. Because of the drought, the teak trees stood grey and leafless. The bush fires around them, that had created such an exotic smell, had no effect on the life of the trees.

I had a contact to make in Chiangmai, a Mr. Prasit. The monk suggested that I spend the night on the stage of his huge Buddhist Assembly Hall, and the following day he would telephone Kun Prasit on my behalf, and at a convenient hour. This was a town where I imagined my monk was extremely well known. So an introductory word from him might be useful. I accepted the invitation, and spent another night on hard boards.

CHAPTER 18

Chiangmai

MARCH 1959

Early on my first morning in Chiangmai, the night having been spent on the floorboards of an Assembly Hall stage, my friendly monk, with whom I have travelled, telephoned Kun Prasit on my behalf (the Kun part, I imagine, is the equivalent of Sir). This grand and very nice person then came to collect me from the Assembly Hall, and I was taken for breakfast in a cinema café where the local dignitaries collect in the cool of the morning before starting their day.

I kept a very low profile as they discussed matters with some solemnity. This gave me a good chance to look around

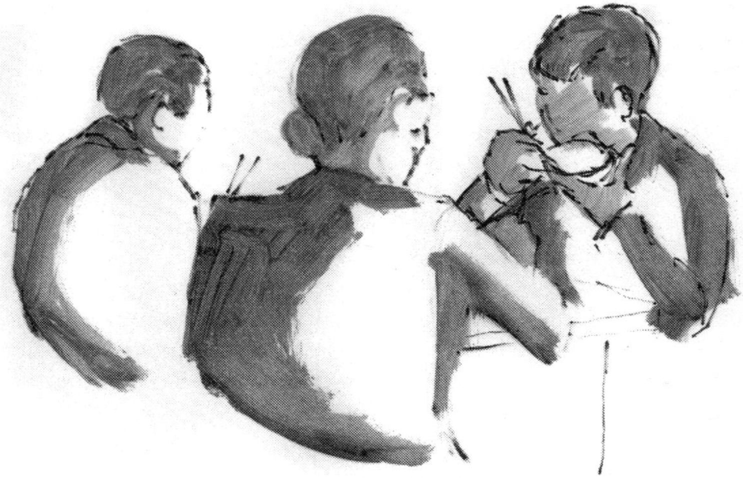

the eating place. Well, it must have been one of the better cafés in which to eat, and the food given to me was fine. But in England, such an establishment would have been given a good fumigating first, and even then given a good berth. But I am getting into the swing of things and have no qualms about hygiene, consuming almost anything with pleasure. After all, if I had not reached my present attitude to eating, I would be starving. My resistance to noxious bacteria must now be considerable. Are we, perhaps, too hygienic in the West?

Anyhow, my luck is in. The outcome of this parley, as far as I am concerned, is that Kun Prasit, who appears to be the top dog hereabouts, and a millionaire, is arranging for me to stay in a desolate, wooden house of his in the mountains. I must provide for myself. The place is among his tea plantations and the hill people. Moreover, another man at the meeting has organised a prospecting journey to the highest mountain of Thailand, Mount Doi Internon. This is near to the Burmese border and will take several days. The object of the safari is to see if it might be possible to build a hotel conference centre in the cool of the mountains. But first, the blessing will have to be obtained from the Meo chief of the mountain tribes. Would I like to go along? I accepted immediately, and with profuse thanks. I must buy provisions for myself, and be ready at crack of dawn tomorrow.

Beside the provisions, I now needed that essential bowl for when eating, drinking, washing or flushing. Kun Prasit's manager, Sum Ta Wee, took me to buy one. The industry for making bowls of the smarter kind is a decentralised one. In the first house that we visited, old silver Chinese coin and Indian rupees (when they were silver) are melted down. In another house, the softening and hammering of the silver takes place. The product now becomes plate-shaped. In yet another house, the silver plates are hammered into a bowl shape and, in yet another, a girl, using a variety of small tools, hammers in the raised design around the rim. The patterns, made up as she goes along, are sometimes most intricate. I have chosen the simplest. It is a work of both art and craft, shows every hammer mark, and is a lovely object.

The paper on which I write is Japanese and, as you see, semi-transparent. It comes with a double-sided line guide to place beneath each sheet. One set of lines is to help the horizontal writers and the other for those who pen in verticals.

I have bought provisions, under Sum Ta Wee's guidance, and am about to spend another night 'on the boards' at the Buddhist Assembly Hall. Jim.

CHAPTER 19

Expedition to Mount Doi Internon

MARCH 1959

Our leader and one other member of our "safari" collected me at six in the morning in a large open truck. It was not long before we had added eight others and many baskets of supplies. The most interesting aspect was that my companions were all armed, most with revolvers in the belt. It looked as if there might be danger ahead. Anyhow, everyone was in great spirits and clearly looking forward to the adventure.

The roads leading in the direction of the Burmese border were very rough. But after the ride north from Bangkok, anything was now acceptable. My spine has taken on new support muscles.

We passed many people walking to market. They were mostly women. Their goods were carried in baskets suspended from either end of a shoulder-carried bamboo pole. It was easy to see what was in these baskets as each was woven specifically to fit the animal inside. There were pig baskets, duck, and chicken baskets.

The women carriers of goods, and women at market stalls, were smoking enormous cigars with banana leaf wrappers. The men smoked a smaller edition.

After about half an hour we reached a place called Chom Tong, where we breakfasted. Placed before us was glutinous rice, vegetable/chilli mixes, meat and fish curries, hard-boiled eggs in spicy gravy, raw cabbage and herbs, one of which was quite bitter. I had a quick look around to see how to eat the food, and copied. With the left hand I pulled a good lump of glutinous rice from the pile on the table. Now, using the right hand, small pieces were pulled from the left-hand supply. These were rolled in the fingers to the shape of a ball, and a depression added. The ball of rice was then put into whatever dish chosen, and a morsel of it pressed into the rice with the thumb. Now it was popped into the mouth. Such a satisfactory method of eating required neither plates nor cutlery. And the food was delicious.

Here we gathered extra porters, to bring our number up to thirteen and, still in our truck, made our way to base camp, which, I think, is called Mae Klang.

In the same way as the market women carried their goods, so did the porters carry our provisions at the ends of bamboo poles. Off we set on foot, with me dressed in a khaki shirt, light trousers, leather shoes, and carrying a water bottle and drawing block. Jim.

CHAPTER 20

Into the Mountains

MARCH 1959

The first part of our climb followed a stream, by which there was a working, overshoot water mill. Obviously I could not stop to investigate. Soon a distant roar of water told our leader that we were about to reach the river we would follow for the rest of the day.

'River' hardly describes this current of water. It rushed over slippery rocks, smoothed by time and coated with slippery algae. It cascaded, tumbled and tossed, fell and splattered, slowed down, rushed, twisted and turned. After the dry and dusty countryside of our approach, this water became our guide and cooling companion for the rest of the day.

Our trail sometimes ran beside it, and sometimes in the jungle. When we had to cross the river, matters became both difficult and hair-raising. For the porters it was nothing, as the crusty soles of their bare feet managed to find grip on slimy rocks beneath the rushing water. For the rest of us it was quite a different matter. After testing several theories, my technique became to take off my shoes, socks and trousers, and try to do as the porters did. But, oh dear, it was a frightening experience whenever a crossing had to be accomplished.

The tracks we were now taking were through small, leafless trees, devoid of animals or birds as far as I could see. But the trees grew larger and greener as we gained altitude. Where the trees were tall, it was noticeable that orchid petals lay in our path and floated on the surface of the rushing water. And there they were, in full flower, high, high up on trees without footholds.

I have begun to take stock of our party and discover their objectives.

The envisaged hotel is to have a service road, helicopter station and dams to supply water and hydro electricity. So to survey for this is one of our main objectives. But, first of all, permission will have to be gained from the chief of the Meo hill tribes. (the word Meo is sometimes written as Maeo or Meaw, but is pronounced 'meaw' — as in a cat's meaw).

Of our number, the leader comes first. He is the ideas man and the power behind the whole expedition. Next comes a sportsman. He is a jolly Thai, who has made much noise on the ascent so far. He laughs a lot and is liable to fire off his gun at any time, scaring the hell out of the rest of us and making us jump. There is a professional marksman, and a student, who, I think, is studying the hydro electric side of things. And there are others, myself, and the porters.

Guns are loaded at all times, and generally seem to be pointing in my direction. Our sportsman did actually aim and fire at a jungle fowl. But as the shot in his gun was meant for elephant, he had little hope of hitting such a fleeting target.

At one crossing of the torrent, the leader, who was directly in front of me, slipped and disappeared beneath the water to be swept downstream, gun, Leica camera and all. I followed him almost immediately, soaking everything and ruining my drawing paper. Fortunately there was another supply in my kit bag, held securely on my porter's bamboo pole. At least, if anyone falls in with my kit, I know that my valuables will be safe and dry, well sealed within the blow-up pillow.

After this incident, my crossing technique changed. Now it was trousers in situ, and Japanese open sandals on feet. But whenever a bridge, in the form of a springy, fallen tree, needed to be negotiated, it was back to bare feet for a better grip.

Even my rudimentary jungle skills told me that we were following an elephant along a narrow track. Its traces nearly blocked our pathway.

At one particularly steep part of our trail, the porters consulted with one another and refused to go further. It was a good-humoured strike, and soon over.

It was always wonderful to stop for a breather, our bodies having been stretched to their limits on the arduous climb. At least there was no dearth of water to drink, with a constant stream of it nearby, running cold and clear.

Lunch was taken, with the rice placed in a bag and kneaded, pushed and pummelled in running water for a long time before it was boiled and served with spicy dishes.

Soon after setting out after lunch, we saw our first elephants. These two were pulling logs on chains down to a dry riverbed. Then we saw another — loose. It was on its own, munching away at forest greenery. It was noticeable that our pace quickened quite a bit as we passed it.

We camped out for the night in the jungle, with me stringing my mosquito net between trees in the undergrowth. Damn the wild creatures around us, we were all far too exhausted to bother about them.

As we gained altitude on the following day, to where pine trees grew, the undergrowth became drier, with much of it on fire or smouldering. The resinous smell was intoxicating. Some of the pine trees had gaping and charred holes at their bases. This was where hill people had gouged out and then burnt into the trunk to form a cup to catch resin. Some trees were burning through where they had fallen to the ground, the remains of others could be seen only as a thick line of ash. It was difficult to know if this was the work of nature, or the clearing of land for opium poppies.

At last we left the river behind, climbed over a steep ridge, and there below us was a panorama of rice fields. In the middle stood a worker's house on stilts. This was to be our resting-place for the night, the residents having moved elsewhere.

To gain access, it was necessary to climb a ladder from beneath and enter through a trap door in the floor. It was not just a shelter from the elements, but also protection from wild animals, snakes, etc. Yet the porters must sleep on the ground outside.

The house, one of a few spaced about half a mile apart, is normally lived in by a rice-growing, hill family. In it they live, cook and keep their agricultural tools. There are digging implements, a bullock-drawn one-furrow plough, and a large comb rake. In the apex of the thatched roof are stored mats for rice winnowing and a long-bladed cutlass. The walls have an inner lining of bamboo, which helps with wind proofing. And, as usual, there are large cracks between the floorboards.

The roof thatch is built up from overlapping layers of bamboo on which rice straw had been bent double and sewn to it — like long tiles. The apex is a V of straw sewn to three poles. But the roof is in such a poor state of repair that much work will have to be done on it before the start of the rainy season. There is no vent for smoke.

The men who work here wear denim jackets and baggy 3/4 length trousers. The women all wear a grey cloth turban folded flat from forehead to nape of neck, and a 3/4 length smock, the lower half of which is embroidered with flowers and horizontal red bands. Beneath is a skirt of ankle length, decorated with red or yellow ochre horizontal bands. They smoke either large, banana leaf wrapped cigars, or small Sherlock Holmes pipes, ornamented with silver.

With a night's break on the way, we have climbed through over 30 kilometres of mountainous jungle to reach this place. So, after a bathe in an icy stream and a good slug of Thai whisky, there has been time to write this. Now it is glorious, glorious rest to come. Jim.

CHAPTER 21

The Chief and Mountain People

MARCH 1959

And I thought I would have a night of blissful sleep and rest after the two day, 30 kilometre climb! Not at all. As darkness fell, it became cold — very cold. Fully dressed, with my feet in the kit bag and wrapped in my sarong, and with every item that might provide warmth around my body, I shivered until morning.

After breakfast of coffee and toast, we added two more porters to our train. Then, after only an hour on the trail, being followed all the way by a bevy of small boys, we were at the foot of Mount Doi Internon. This was to be our final camp. Accommodation was to be in a stilted, thatched hut, beneath an enormous clump of bamboo beside a stream. Dry paddy fields surround us.

Plans are to spend one day surveying and reconnoitring, and the next to make an assault and return from the mountain.

I spent the morning drawing. And after lunch and siesta, the chief of the local hill tribe appeared for the conference, bearing a gift of two bottles of violent spirit — real firewater.

At the meeting he was assured that his tribe would not suffer from the proposed project, and that by changing from growing opium to coffee, tea and fruit, they would all be better off financially.

It seems that during this parley he said that as he lived nearby, he would gladly escort me to his village. In trying to estimate how long I might have to draw in the village, I enquired how far away it might be. The answer came that it was about as far as we had travelled that day, namely an hour's

trek. Which sounded fine. So one of our porters was provided to guide me back, and off we set with the chief in front.

The chief is a finely featured man of modest age for his exalted status. Of medium height, his hair has been shaved off, except for a circlet at the back of the crown. From this grows a length of jet-black hair, plaited into a single rope. About his neck he wears two large silver necklets. On his chest he wears a black, double-breasted bodice, buttoned under one arm with a silver button. The front flap of the bodice is decorated at the edge with embroidery. Then there is a gap of brown skin between the bodice, and calf-length, blue-black, baggy trousers, secured by a scarlet, knotted cummerbund. On his feet are brown sandals without laces. The outfit is completed by a double-pouched, embroidered bag, slung over the shoulders. He cuts a fine figure.

I carried a tin of biscuits as a gift and a pocket-full of boiled sweets, as well as my drawing paper. The return guide carried a bottle of tea.

The route we took followed an indistinctly-defined trail, and was precipitous to say the least.

After about half an hour's travel we came across an opium patch. This took the form of about an acre of cleared jungle, the trees growing there before, having been cut off at the height of about four feet. A rough fence of prickly limbs protected the patch from intruders. The season was over and only tall, dried poppy seed heads remained. I resolved to gain entry on my return to harvest some seeds to grow in England.

Quite naturally, I imagined that human habitation was near at hand and that there was not far to go before we reached the chief's village. It was very hot and I was quite tired from scrambling up through jungle, sometimes having to climb by pulling myself up on the roots and branches of trees. But we had only covered a quarter of the distance. It was worth noting, that although we only took three rests, the chief and my guide were as glad of them as I was - and they were sweating more profusely.

We passed through burning patches of forest where fallen trees were burning to ash.

On reaching the foot of a particularly precipitous ridge, I raised my eyes from seeking footholds to find in front of me what at first appeared to be a one-bar fence, supported by a series of elongated catapult handles. Then I discovered that the one bar was bamboo sliced in half lengthways. Overlapping as it went, it carried a constant flow of clear water. We followed this aqueduct for about half a mile around the side of a very steep slope. There was hardly a sign of leakage from this fine piece of engineering.

We then passed through three thatched arches with bench seats on either side. The significance of these I never did discover.

Then, suddenly, the whole nature of the ridge changed. Before me, on a dried mud plateau, were twelve, thatched, bamboo houses, built on either side of a main thoroughfare.

Pigs, dogs, goats and chickens were all about, and as the village first came into view I noticed one or two women and children, just standing idly about near to their homes.

One woman was collecting water from the ever-full bucket at the end of the bamboo water supply.

The chief lived at the far end of the left-hand row of houses and, accordingly, we marched down the street. Some people ran into their houses when they saw me. Others stared blankly. One smiled. The door of the chief's house was closed. He opened it to reveal two old ladies, one woman feeding a three-year-old child at the breast, and three other children.

The women were dressed alike. They each wore a turret-like, blue-black hat, a loose jacket, buttoned low to reveal most of their breasts, and a long, what appeared to be a trouser-skirt to ankle length, with scarlet waistband. Each wore silver necklets and chains, and silver earrings. None wore shoes.

Their hair stuck out in straggly fashion from under the tall hats. Their skin was clear and yellow, and even in old age, for the most part, without lines. All were round faced with features clearly defined. Each was ready to smile and laugh.

One of the old ladies was smoking tobacco from a bamboo pipe. This pipe was not unlike the opium pipes in the Nakorn Sawan den. It was made from about 2' of bamboo, with a diameter of about two inches. Quite open at the drawing end, it was bound along its length by several wide, silver bands. Some one third of the way up from the tobacco end was a small diameter projection, again of bamboo. This pointed upward and outward. Into the end of it was placed about a thumbnail's worth of chopped green tobacco, taken from a silver, rectangular tin. A wooden ember was then taken from the fire to ignite it. The bamboo

contained water. So when the limit of three or four puffs were drawn up, loud gurgling sounds issued forth. Then, after a quick blow, a back-spurt of dark liquid cleared the bowl, and more noises followed, even when the pipe was put away.

The girl children's hair was uncut and unkempt. The older one wore clothes like the women, but the younger wore a long jacket that was either open or closed. They adorned themselves with the same kind of ornaments as the women, but without earrings. The boy wore only loose trousers, held at the waist by a scarlet band. Otherwise, he had a silver

necklet and chain around his neck. A catapult and a tiger's tooth tied to a dark cord completed his ensemble.

The house had several fireplaces but no visible outlet for smoke. One fire was alight.

Furniture consisted of shelves, blocks of wood to sit on, and one four-legged stool on which sat the feeding mother. The floor was of very uneven, dried mud, with a dusty top layer. In the house lived pigs and chickens, and I gather the goats, too, came in for the night.

There were quite a few black mountain dogs, resembling leaner and less furry chows. The pigs looked like wild ones. There was no sign of disease among people or animals.

After some drawing and sweet-giving in the chief's house, it was indicated that might I like to see another house? I was taken to one where a young, rosy-cheeked girl retreated in haste upon seeing me. Then a young man appeared, dressed like the chief, but with uncut hair. After a consultation between them, it was decided that I should go to another house. As we approached this one, sounds of chanting, accompanied by regular beats of a gong, issued forth from the doorway. Inside was a veiled man, dressed in long black robes, which were tied in with the traditional scarlet waistband. His head was decorated in tatters of black and red material. He was

sitting on a cushioned seat in front of an altar, wailing and chanting and singing, and bouncing up and down on the seat, rising each time by putting pressure in alternate feet. Behind him was a tall girl beating a gong, and in time with this, the old man was wildly shaking a large diameter silver ring, on which were strung pieces that looked and sounded like cymbals. I was the only one taking any interest in these proceedings.

The altar took the form of shelves adorned with cut-out pieces of paper, rather as if the paper had been folded many times and cut to flower and star shapes before being opened out. On the central shelf stood a bowl, containing burnt and burning incense sticks.

The chanting continued at a very fast rhythm for the time I was there. Once or twice, however, the chanting slowed down and became a kind of speech, unwinding to a stop, like a gramophone needing to be wound up again. When this happened, the girl with the gong accompanying the chanter, held his shoulders to prevent him from toppling over, that was, until the human gramophone slowly started up again, soon to reach its fast rhythm once more.

This house was roughly the same shape as the chief's, but two fires were burning, and two people were smoking pipes.

I saw no bed space or bedding. Only chickens were visible in this house. And I don't believe that washing the body was customary.

Because of jungle hazards and the approach of darkness, it was necessary to leave. Although I have now become quite used to eating in unhygienic conditions and sleeping rough, I would not have been comfortable had I been invited to eat and stay the night in this village. But they were delightful and colourful people.

On leaving the village, a bearded young man with uncut hair appeared from a house. He would not have looked out of place on the King's Road in London.

It seemed that word had gone around that I was leaving, as all the inhabitants came to see me go. But my farewell waving meant nothing to them at all. The only returned greeting was from the chief, in the Thai way of hands together as if in prayer, accompanied by a slight bow.

The journey back was not easy, as the jungle tracks seemed so often to cross, converge and diverge. However, the marks made on the outward journey by my Japanese sandals in the dust were a great help.

We passed the opium patch, seen on the outward trek, and I indicated that I would like to go in to harvest some poppy seeds. But, although it appeared to be protected by a few spiky branches, it was almost impenetrable. However, after some difficulty, and not without a few scratches, I obtained my seeds.

About half way back, on a steep downward slope, which incidentally was covered by spiky nut husks that often had to be extracted from my guide's bare feet, we heard the clunking sound of a wooden, cattle bell. I imagined that we might see a buffalo with its attendant. My guide, who was in front, pointed at something and retreated at great speed. There, in front of us, were three large elephants eating jungle greenery.

We had approached from downwind and it was not until we were quite close that we saw them, and they scented us. When the one nearest opened wide his ears, it was our cue to retreat. This we did very quickly.

My guide, who obviously knew more than I did about such matters, was all for a grand detour, but since our tracking was not of the first order, I didn't want to divert too far from the trail. We could not afford the time to wait, as it was absolutely essential to get back before dark. A compromise was worked out, which involved a slight detour around the elephants. They then crossed the path behind us. We moved slowly on the detour, speeding up no end having passed them. The sound of the wooden bell travelled a great distance in the jungle, so it was difficult to ascertain how far they were behind us.

We were somewhat nervous as we travelled onwards. And it did occur to me that here were two jungle simpletons, far from our fellows who had thought it necessary to arm themselves with everything from revolvers to small bore rifles and an elephant gun. What might lurk around us as it became darker? So to break the tension, I let out a deep-throated growl. The effect on my companion in front was electric. He had clearly been thinking on the same lines, and almost jumped out of his skin. From then on we were both laughing and growling for the rest of the journey. Jim.

After this adventure to see the mountain village home of the Meo chief and his people, I was almost unable to stand up for exhaustion and nervous tension.

We had not met another person on the climb or descent, and the only cultivation witnessed was that of the opium patch. There was no cultivation to be seen either in or near the village.

When darkness did fall, the jungle about us seemed to be in full flame. Yet, on the following morning in daylight, there was no sign of burning to be seen from our hut.

With parleying done, the day was to be taken up by an assault on the mountain by my companions. I had done my bit already, and there was no artistic bait on offer this time. So, for me, it was to be a day of catching up on drawing, painting, writing, and regaining strength. Leg strength was especially important, for on the following morning at daybreak, we set off on our return journey to base and homes.

This journey involved getting lost several times, some of us nearly falling down a dusty precipice, our sportsman trying to shoot a rather lethargic-looking wild boar with a rifle that didn't work (and anyhow, the barrel had been blocked by earth when he had used it as a walking stick), a stay with hill people who had graduated to denim dress, a shared bus seat with two boys and a bag of very black and dusty charcoal, and another bus ride with me being drooled over by two cows that had been separated from the passengers by three bamboo poles.

I established myself in an hotel in Chiangmai and decorated some pottery for one of our safari number who wanted to commemorate the escapade.

When eating at a café, a complete stranger came up to me and paid for an expensive dish of buffalo meat. He had seen me on the train going northward.

I was then taken to Sun Kampeng to meet and draw the not so pretty 'Queen of the Silk Industry' and on to see ebony elephants carved with great facility when held between the toes of sculptors. Then a stranger gave me a silk scarf for 'after tennis'.

CHAPTER 22

Some Animal Life in Chiangmai

APRIL 1959

I have had my first encounter with a little lizard (a gecko). He could not have done his rightful work in helping to keep my room free of insect life (even though I have seen nothing penetrate the protective netting in the window or enter through the doorway), because of where he lived.

He had selected as his den the inside of the lip of the lavatory basin in my room, and never ventured forth. Every time the lavatory was flushed, the little fellow had to swim back up the waterfall with all his might to regain his chosen territory.

Naturally I felt very sorry for him and worried about his future every time I came to flush. Then there was the slight worry that I might be bitten on the behind. But he was a charming fellow, and that was the risk I took. The risk he took was far greater.

So I thought that he might like to leave his lair for a safer one of the floor, walls or ceiling of the room. So I made a bridge of coat hangers from the lavatory basin, over the seat, and down to the floor, and baited the trail with tasty titbits.

But presumably, as he was a drain gecko and not a room gecko, my offer was declined. Then, after a particularly strong cascade of water, he was flushed off to spend his days somewhere downstream — where I hoped the fare would be more appetising and varied.

I have changed hotels. When looking at my map of the world it is very difficult to gauge, with any degree of accuracy, how far I can go and still get home on the money in hand. So this change to a cheaper establishment is a precautionary measure.

Immediately I entered my room in the new hotel in the evening I saw him, and was quite taken aback. He is pale grey in colour and decorated with pink and white spots. That the room belongs to him there is no doubt. It's just that will he let me stay?

He is a giant house lizard, presumably a gecko, of about 18" in length. If I am the guest in his room, as I have decided to be, at least I am larger by quite a bit.

I made enquiries about him and was told that he was very useful, eating anything from cockroaches to rats.

His daytime home is behind the wardrobe, hidden out of sight. I would like to make friends, but I am certainly not going to dangle my hand behind the wardrobe to flush him out for a chat.

The room is netted and, as far as I can see, there is no food at all for him to eat.

At night he comes out and runs all over the walls and ceiling, defying the forces of gravity.

We gained confidence in each other as the days and nights passed — that was, until I rose one night to cross the room for water. In getting back into bed in the dark, he fell right on top of me from above, giving us both a considerable shock. We now treat each other with more than a little apprehension.

I noticed that several times during the night, a possible mate comes to the outside of the window netting. So I resolved to be a matchmaker of lizards (why I try to muck around with nature I don't know).

So one afternoon, when the mosquito population was asleep, I pulled the hinged frame of net in the window ajar, and eased the wardrobe away from the wall. Now he would be able to escape and lead a much fuller life in the great outside. He would not contemplate it. So I dangled the leather strap of my water bottle between wardrobe and wall to encourage him. He rushed out into the open with glaring, orange eyes to threaten me, uttered a sort of croak/growl, and went back behind the wardrobe.

I took the chance of getting bitten to death by mosquitoes and left the gauze window frame ajar all night. I'm sure he didn't venture out, and I don't believe that another lizard ventured in. So the status quo remains, and I'm absolutely sure he'll be extremely glad when this meddlesome Englishman leaves on his travels once more. Jim.

CHAPTER 23

Some Lessiu People

Chiangmai stands about a thousand feet high on a plain at the base of high hills. I was invited to visit these hills by Kun Prasit.

Kun Prasit, my mentor in Chiangmai, is a delightful and worldly person with clearly great business acumen, power and wealth. Moreover, he is a benefactor for rightful causes, and a crusader for the betterment of Thailand, and especially of those who live in his part of it. Although extremely busy, he took me to the hills in his Land Rover.

We climbed another two thousand feet toward a monastery on top of one particular hill to see a friend of his who lived nearby. The friend was not there, so we sat beneath huge evergreen trees to talk. As we did so, a vivid red bird flew from a patch of sun into shade. And all the time, in the background, exotic noises came from the forest.

Whereas the jungle in Malaya was tropical jungle, here it was arid jungle, where trees grew tall and leafy only at altitude. Being high enough where we were, the trees stood tall and magnificent above us, with the canopy providing shade from a great height. Orchids thereabouts were rare, he told me, and found only far from human tracks. But I had already seen plenty on my safari to Mount Doi Internon.

Among his many interests were tea, coffee and fruit plantations. It was his aim (I believe self-appointed) to try to establish permanent settlements for the nomadic people of the mountains. One reason for this was simply to preserve the forests by preventing erosion from denuded high land. The tribes, in their nomadic wanderings, destroyed large tracts of jungle to plant rice for themselves and profitable opium. Then the rains would come and wash away the soil. Another reason was to establish national friendship with the shy and potentially dangerous people on

Thailand's volatile northern border.

Then there was opium. It was against the law to grow opium in Thailand and had been for many years. And in two month's time, from when we spoke, it would be against the law to smoke it, even though the government had been making a great deal of money from the dens, which were under their monopolistic control.

That was all clear, but the people of the hills had grown poppies and harvested opium for ages. It was their only form of livelihood. Kun Prasit's task was to encourage these people to grow coffee and tea instead, and thus provide them with a lawful income.

This task was made slightly easier than might be immediately apparent, as the people had always been in fear of the police, who had bled them with graft. And to live no longer in fear of the corrupt police force was, in itself, a considerable inducement to abandon poppies for tea.

In the course of learning about opium and the hill people, Kun Prasit told me that the opium I had smoked in Nakorn Sawan had been very weak, as the government had purposefully reduced its strength, with prohibition in view.

The friend that we had gone to meet still did not appear, so we headed for base. As we travelled I noticed that although Jeeps and Land Rovers were the most popular vehicles in the hills, Japanese Toyotas were in the ascendancy on the plains. Toyota was also exporting many commercial vehicles to Thailand.

We stopped for food and drink, and I could hardly help but notice that spitting was a universal habit. Spittoons were often provided in eating-places but not always used. And the air was rent with ear-splitting throat clearings.

On our return to Chiangmai, I was handed over to Sum Ta Wee, Kun Prasit's secretary/manager. This man had business to conduct in one of the mountain tea plantations. Would I like to join him and his family for the visit? Here I would see an example of the rehabilitation of people who once grew opium. You bet I would.

As I would be staying there by myself, it would be necessary for me to buy provisions. So, to the normal contents of my kit bag, I added some tins, an opener, boiled sweets for children and a bottle of Thai whisky (costing the same as a bottle of fizzy orange drink). There would be fruit aplenty available in the mountains, and water flowing in the streams.

With the family, we travelled over the usual rough roads, making a detour to visit some Buddhist caves. Camping outside them was a troupe of Lessiu hill people. They were opium smokers and producers, going deeper into the mountains to ply their trade now that both the growing and smoking of opium were to be forbidden.

Lessiu men wore the ordinary denim jacket of the countryside, but with a light blue or green top skirt over black leggings with red turn-ups. The women were a dazzling sight. Above the waist they wore several overlapping, embroidered garments with multicoloured hooped sleeves. Below were dark underskirts with bright blue or green overskirts. Long red tassels hung down from their waists. Below the skirts and above their bare feet, they wore black gaiters, decorated with a red diamond pattern with scarlet turn-ups. To top off this most colourful of costumes were enormous plate-shaped headdresses.

Both men and women chewed betel nut (the seed of the betel palm, wrapped in betel leaf and lime), which gave them blotchy, scarlet mouths. The old ladies presented a particularly grotesque appearance, as red saliva ran down from their mouths into the natural channels created by wrinkled skin.

Being a shy and retiring people, we did not intrude in any way.

The crops growing in this area were varied. Soya beans were growing in the rice fields, as the rice had been harvested. There was tobacco, cabbage, tea, tangerines, bananas, papaya and garlic, the last being used in great quantity in both Thai and Chinese cooking.

The houses all had generous provision for rice storage. Bins for this took the form of either top-heavy-looking barns on stilts, or roofed-over, large, woven, mud-plastered baskets resting on platforms. They, and the houses, formed the main features of the landscape.

CHAPTER 24

The Mountain Tea Plantation

March 1959

With Sum Ta Wee and his family, I set off for Kun Prasit's tea plantations. At checkpoints where the mountains joined the plains, all vehicles were stopped by police. But only our faces were searched, presumably because regular traffickers were known, and amateurs would give themselves away by looks of guilt.

I am writing this in Kun Prasit's simple bungalow overlooking hills of tea bushes and valleys of tangerine trees. It is lovely.

I have been thinking of my next move forward. To rush on to Japan for the cherry blossom season would now be foolish. It is probably over by now. South-East Asia is so interesting that I would be silly, I think, not to stay for a while. Looking at my map of the world, Cambodia, with its Angkor ruins, seems to be comparatively close, and Vietnam not so far beyond.

As I was writing that, an ancient horse/mule creature appeared at my doorstep, coaxed by a muleteer and followed by an attendant. The mule was for me to ride, the muleteer was to drag us both forward, and the attendant was to carry my baggage. Life's full of surprises.

The mule did not care for its task and was unsure of its footing, even on level ground. When we came to narrow and unguarded mountain tracks with precipices below, and with my bottom saddle-sore, I opted to walk behind the beast.

This now gave me a chance to look around rather than spend my time holding on for grim death. There were trees that grew separately, then to join together before becoming one single tree, rising upward into the jungle canopy. Hanging from branches were pendulous, crescent-shaped, mud beehives, with many flight holes. There were streams with orchid petals floating on the surface of the water, and bee-eater birds that climbed steeply into the hot air making lots of noise, only to come tumbling down, seemingly out of control, to correct at the last second and land safely on a branch.

After negotiating many kilometres of mountain trail, we came upon Kun Prasit's isolated tea plantation for hill people. It had been carved out of virgin jungle and terraced. The tea bushes were stunted and plainly having a struggle to survive in the hot and dry season. But more jungle was being cleared, and the timber stacked for drying tea. The tea made here is dusty, Indian style, and green, Chinese style.

The house I have been given rests on hardwood stilts, but is otherwise constructed entirely of bamboo. And for this time of the year it is ideal, being shady, breezy and animal-proof. There is a balcony, a main room, and an ablutions closet. It has all the charm and fun found in a child's tree house. For the rainy season it would be a disaster, with its roof in tatters.

When contemplating the delights of my new abode, a sprung bed (glorious luxury) was brought in, with a pot of green tea. The providers of these items (who originally had to break the door lock for me to gain entry to the house) are Chinese looking with strong and prominent jaws.

The person in charge has come to see me. He is clearly a man of intelligence, and is quite fascinated by my Thai/English phrase book. Two of the items that made him smile, were: "I want a woman" and "Can't you make it any cheaper". The two went together. I somehow imagine that those are among the most used phrases by the male and lusty foreigner. To make myself understood is somehow far easier with more complicated matters than with simple ones. But making drawings is the best, most successful, and by far the most enjoyed method of communication. For instance, I drew a bunch of bananas and handed over 5 bt to a villager. Back came a hand of 30 bananas and 3 1/2 bt change. The fruit cost the equivalent of 2 1/2 pence - nothing.

The bananas came from a Meo community nearby, whose people are looking for a permanent village site. They are camping at the edge of very thick jungle in lean-to shelters made of sticks covered by leaves that are gathered wet and flexible. Red blankets, coloured cloth, and a mat or two are laid out on the ground or on the roofs. Without this colour, the encampment would be almost invisible, so well does it blend in with the jungle foliage. But the atmosphere in the little community is one of sadness and decline.

When I first came across them, the women were out looking for roots, while four men were smoking opium in one of the shelters, using the same kind of pipe seen in the Nakorn Sawan opium den. They sat around in a half drugged, expressionless state. When all the tribe were together, I noticed that many were suffering from some form of cold, with runny noses, sneezing fits and watery eyes. Several coughed in a tubercular manner.

P.R

The women, whose breasts, with extended nipples, hung through the opening of their jackets, wore the same garb of the Meo seen in the Mount Doi Internon village, except for wearing black gaiters, being hatless, and with hair piled up in buns.

One young man, with a baby strapped to his back, held a musical instrument that looked not unlike a decorated table leg, from which rose five, pierced bamboo projections. Although very shy, I eventually gained his confidence enough (through gifts of sweets) to get him to play it. It took a great deal of blowing. Instead of the expected flute-like noise, the sound of an organ issued forth. After a very short recital he was quite out of breath.

A Mussir tribe had built a village on a dry slope not far away in the jungle. The inhabitants were less colourful than the Meo, with the women and children wearing towel turbans on their heads. A man and a boy were operating a rice-husking machine, which took the form of a giant hammer on a fulcrum, with the head falling into a conical bowl. As they were operating this, a sow and a boar started to fight. The boar tore many nasty deep gashes in the sow's body. Yet no-one took any notice whatsoever, despite the fury of the fight and the blood-letting.

The Mussir tribe were doing much of the tree cutting and harvesting of wood for Kun Prasit's tea plantation. A few of their strong men stayed at home with the elderly, but the rest worked during the day. They appeared to be much healthier than the Meo, and obviously more hardworking.

Around mid-day, when the hot sun beats down without mercy, the noise of children playing resounds about the terraces as if this was the seashore. My sweets have been a huge success in making contact with mothers and children here. And my bottle of Thai whisky has also been invaluable in the social context.

'Chews' are popular with all. Besides the betel nut chews that make the mouth and lips look as if they are bleeding, there are others. Of these, the nicest is made of tealeaves, ginger and rock salt. Tealeaves are soaked or boiled and

squished into a large ball. From this a few wet leaves are extracted and into them placed thin slivers of gingerroot and rock salt. This is then folded into a ball and chewed. The waste is spat out.

The large cigars smoked are simply made. Green tobacco is cut finely, using leaf, stem and all. Then it is left in thin layers to dry in the sun. After that it is just rolled into a piece of banana leaf to the size desired, and tied around with a thin piece of material taken from a wodge of cotton waste. It has to be re-lit quite often. It is amazing to see young girls and boys smoking these vast objects.

A very amusing incident occurred as I was drawing. A handsome and tall Mussir woman appeared from the jungle, dressed in full regalia, with possibly a three-year-old child feeding at the breast as she walked. She was smoking one of the gigantic, banana-leaf-wrapped cigars. I offered her boiled sweets. She took one, crunched it up in her mouth, took the child from her nipple, gave it some crunched-up sweet, and then, would you believe it, put the cigar into the child's mouth for it to enjoy a puff or two.

I had drawn a Meo man in the tea plantation, and he has followed me back to my hut to watch me record the happenings of the day. So here the two of us sit. He is wearing a black skullcap with magenta tufts. There is a pigtail hanging down behind it. On his top he wears a black waistcoat, decorated with orange and white flower patterns and from which hang multicoloured thongs. It is held together with a button that is also a bell. Baggy, 3/4 length black trousers are adorned with white piping, and on his feet are brown, laceless, sand shoes. I am wondering if I look as strange to him as he does to me. He is a young man, and only sits, stares and yawns.

When the women returned to the encampment with full baskets of roots, the atmosphere changed immediately. They now helped to care for the children and cooked rice and roots in bamboo tubes for their families, pigs and chickens. They sat in rows outside their abodes like contented birds on a perch. Surprisingly, one mother had two small children with fair hair like mine. So there was much pointing at me and laughing. Although two old women sat and smoked tobacco in an 'opium' pipe, there were no old men to be seen. And no-one was smoking opium in the hut.

Opium seems to have little effect on the men's virility. One young man had four nursing mothers under his care, and looked particularly happy about it. And nearly all the women carried a baby on their back, slung in a cloth.

Although I did once see a crossbow in full working order, these men owned guns that might have been acquired from Oliver Cromwell's Army Surplus Stores. I had a chance to hold one of the muzzle-loaders. It was extremely light and easy to handle, with a barrel twice as long as a modern shotgun. As a flintlock, it had a small piece of material lodged between the hammer flint and ignition tray to keep the gunpowder in place. The charge was a coarse, granular mixture, carried in a blue, glass medicine bottle. The roughly spherical lead balls used as shot were home-made. The ramrod was incorporated into the stock. And the noise, when I first heard one of these guns fired in the jungle, was like a cannon's roar — most impressive. Jim.

One morning a mule train appeared with fuel oil supplies. Sum Ta Wee came with it as the paymaster for those in the employ of Kun Prasit. I was to return on the same mule that so laboriously and unwillingly brought me to the plantation. Now homeward bound, it careered down the mountainside. I was very quick to dismount.

By the time transport arrived late at our base camp, my provisions had been exhausted, except for what remained of those fine bananas.

On the truck journey back to Chiangmai, we called in at a tobacco plantation where curing and sorting was in progress. The tobacco was harvested and cured in a barn, heated by wood fires. The leaves were then sorted and stacked. Sometimes the raw leaves were cut finely by a knife blade as they were pulled through a hole in a piece of wood. This cut tobacco was sun dried for a few days. Then it was ready to smoke. Growing with the tobacco were watermelons and tomatoes.

Here was a far more elaborate water distribution system than that seen in the Meo chief's village on Mount Doi Internon. Undershoot water wheels of 20' diameter, were turned by water released from a dam constructed of bamboo. These lifted small cups of water to hollow bamboo ducts, with tow seals, that fed the precious liquid to all the houses around, and crops.

Thai boxing at a fair was in progress as darkness fell. With the arena enclosed in a woven leaf fence, the boxers, dressed in gloves, shorts and genital guards, prayed for luck, and then punched, hit, kicked, kneed and elbowed their opponent to the beat of a drum. The third of five rounds was in progress when we arrived. The most effective move seemed to be to pull the opponent's head downward and knee him in the face. Action stopped temporarily when both contestants fell to the ground, and ended shortly after when one was knocked unconscious by a right and left knee to the chin. The prize was £1.

Nearby was a theatrical troupe. The actors lay asleep on their portable stage in front of a painted backdrop that could have represented an English country lane. The whole scene might have come from an 18th century drawing. I would like to have stayed to watch the play, but we had to go.

I was returned to my Chiangmai room, with its resident lizard and notice on the wall saying: "Not to bring ladies who are not our room customers into the bedroom". Thanking all my mentors, and allowing them to choose drawings, I made for the station.

On the twenty hour train journey back to Bangkok, where much of the trip seemed to be taken up by removing sleeping heads from my body, I made my decision. If the local fighting on the Thai/Cambodia border was not too intense (as I had been told was the case), I would go overland from Bangkok to see Angkor in Cambodia, and then on to Saigon, in Vietnam.

CHAPTER 25

To Cambodia and the Angkor Ruins

26 March 1959

The only reason that I have discovered the date as I start this letter, is because I have just obtained a ticket and visas for Cambodia and Vietnam.

I was told that, because of border troubles, I would be foolish to travel into Cambodia by any other transport than aeroplane. So I took what I am sure was good advice and booked on a flight through to Vietnam, via Siemriep in Cambodia.

The DC3 Dakota aeroplane is by now rather long in the tooth and fatigued, so it was a surprise to see that I was to be a passenger in one of them. It was also a pleasure, as I never flew in one during the war. I do not know if it was a comfort or not to see that a few extra bolts had been added to the wing. Once the engines had been started, a screw above my head, one that was admittedly only there to do a minor job, started to tighten and then loosen itself on its own accord, with regularity. Then there was a disconcerting thumping noise coming from somewhere under my seat. Anyhow, we took off without mishap.

Below us were the rice plains of Thailand, no roads, canals bordered by trees, and some isolated houses. When over Cambodia, the land beneath was covered by jungle, and houses with their plots of cultivation grouped closely together.

After landing on the airstrip at Siemriep, it was noticeable that plots and houses had fence protection. And the roads were far smoother than in Thailand. This was due, presumably, to the French influence. In fact, Frenchness pervaded everything, with nearly all the tourists being of that nationality.

As the time of my arrival coincided with Easter, and also a visit from the Queen of Cambodia, there were no rooms to be had in the only available hotel (the one other having been commandeered by the military). But a young French diplomat, on a visit from Laos, very kindly offered to share his room with me. Because I was paying my way, we became friends. He had no time at all for the many 'English beggar travellers'. At night he was sharing a bed elsewhere in the hotel with his brother's Cambodian girl friend (who turned out to have contracted a venereal disease, which rather shocked him), so I had the room to myself. We made excellent 'ruins' companions during the day. And it was the ruins of Angkor Wat that I had come to see.

I was surprised to discover that Angkor Wat, although the most immediately impressive ruin, was only one of many of almost equal archaeological importance in the region. Some of these were being prised from the roots and tendrils of powerful jungle growth. And it was thought that there were many temples still hidden.

The central ruins at Bayon, within the original square-moated, Khmer city of Angkor Thom, was a bus ride away from our hotel. So we took this mode of transport north, and stopped at Angkor Wat on the way.

This edifice, and a magnificent one it turned out to be, gave the impression of having been carved from solid rock. But in fact, the massive, carved stones of its construction are laid one on top of another without the use of any binding material. They form the walls of small but lofty palace rooms within towers. The small roof span inside is determined by the limitations dictated by the stones closing toward the apex. The Khmers had not discovered the rounded arch in AD 860-900.

The sculptures of gods, men and animals, and bas-reliefs of the daily life of Khmer people (especially at Bayon), covered all surfaces and were conceived on a magnificent

scale. The skill and imagination exhibited was quite amazing. There seemed to me to be a strong Indian feeling about the work.

Deep in the jungle lay the exquisite and ornate ruins of Banteay Srei, which, in its wonderful state of preservation, was a miniature Angkor Wat, but this time, created with feminine delicacy.

We travelled by bus, shared Land Rover and, when the girl friend was with us, a hired car. So we saw more and more ruins, all created by successive monarchs as their palaces and retreats. The sites were everywhere to be seen. Hundreds could have been visited by enthusiasts with enough time and strength.

Angkor Wat was floodlit at night for a royal visit by the Queen of Cambodia and her retinue. There was to be a theatrical performance and firework display. Flares were lit to illuminate the facade of the building, and impressive it was, until the wind changed and the crowd of us well-wishers were smothered in dense smoke. Then it rained, and the queen left, suddenly, and understandably.

Today has been clear and hot, and ideal for a closer inspection of Angkor Wat. The three of us crossed the two long approach causeways, but were prevented from entering the building because the Queen was visiting. However, we eventually discovered an unguarded entrance.

The massive scale was both royal and religious. In the small, tall rooms sit or stand carved deities. We climbed to the top of the main tower, an exercise needing both courage and rock-climbing skills as the stone pathways were steep and narrow with no guarding rails. With the Queen's music rising through the hot and humid air from below, we felt close to the gods.

Escaping unseen to take up station at the exit to the causeways, we awaited the royal departure. The queen appeared in a sedan chair beneath a large, green silk, Cambodian-style umbrella. Disembarking from this litter, it was made known that she would speak to her people. Whereupon, some 200 surged forward, falling to their knees and clasping their hands together. Her words were well received.

The Queen wore a white top and purple skirt folded around, in the Cambodian manner, with its forward flap drawn between her legs and tucked into a belt behind. So her knees were hidden from the front but seen from the rear. She left to the sound of much cheering. Then followed ex-King/Head of State/Prince Sihanouk, to more cheering and clapping of hands. Now came a General, who made a point of fraternising with his officers. But there was more entertainment in the offing. My diplomat friend immediately hired a very strange-looking local contraption. This resembled a tall, three-wheeled, motorised perambulator. We jumped aboard and made off down the regimented lines of political and military troops. He sported a black monocle, made from the lens of a pair of broken sunglasses, and I wore my hat of some antiquity and distinction. With royal waving we elicited far more cheering and merriment than that accorded to the recently departed rulers and royalty.

Although nearly everyone here is French, there are a few

Cambodians to be seen. Their faces have broader features than those seen in Thailand. They are conspicuous, inasmuch as it is they who sell trinkets, such as knives, flutes and crossbows to us tourists. I was interested in the crossbows and their degree of accuracy when used. So, fixing a bunch of worthless currency notes to a tree trunk, started a competition whereby the notes were the prize for whoever was first to shoot a bolt into the target. The crossbows turned out to be powerful, but inaccurate. The fun engendered by this competition in the jungle, with monkeys swinging in the trees above, made it a memorable sporting occasion for all involved.

There is to be more entertainment on the stage at Angkor Wat tonight. There will be classical dancing. And as the dancers here are world-famous, it should be a wonderful evening.

Despite heavy rain, we entered the complex, only to sit on wet earth and look at enormous, pendulous pockets of water forming in the canvas of stage roofing. The rain stopped and, with the help of long sticks, the pockets of water were decanted on to the ground in front of the stage. Then music from xylophone and drum set the scene for an evening of elegance and beauty.

First, some fabulously attired young female lovelies gave an elegant dancing display of the hand, finger, foot and toe variety. Then a man, dressed as a sailor, danced a simple and graceful nautical routine. This act was followed by one of more classical dancing, before several very homosexual-looking dancers took the stage.

The 'queen' of the troupe, who wore a very realistic-looking bosom beneath his female's top, extracted a breast and sucked the nipple — roars of laughter. Back it went. Then an actor prized out the other breast and sucked it, to the evident show of pleasure by its owner — more roars of laughter. Now the partner held a nipple in his teeth and dragged the owner of it around the stage — gales and shrieks of laughter.

And so the evening progressed, with classical interspersed with the bawdy. When it ended at last, and we were able to lift our bottoms off the wet earth and go home, everyone had had a splendid time.

I never thought I would enjoy ruins. But these are quite exceptional. However, I have had enough, even of Angkor. Tomorrow I continue on my booked flight to Saigon, to see Jeremy Hodson, an American pal from art days in post-war Paris.

In another old DC3 aeroplane, and not having done any low flying since the war, it was somewhat of a shock when, soon after take-off, the pilot threw the old crate into a steep turn and buzzed his girl friend's house. Then we headed off for Saigon in a sedate manner, which was more seemly for such an old aeroplane.

When we started to reduce height for the approach to Saigon Airport, we found ourselves in dense fog. As the only navigational aid on these old kites was a loop aerial, the prospects of a successful landing seemed to be on the slim side. The pilot could obviously not see the ground where it should be, so overshot the field and tried to land once more. Those of us who knew about flying were on the edge of our seats with apprehension. Then he touched down successfully in a thick pea-souper. It was a brilliant piece of flying. But it did occur to me that he might have had a row with that girl friend of his outside Siemriep, and felt that life was no longer of much value. Jim.

CHAPTER 26

Saigon

2 April 1959

My immediate impressions and experiences of Saigon are these: I have eaten Consommé froid, crabe farcie and Yoghurt. So the food is French and wonderful. The shops could be in Paris. The boulevards are wide and tree lined. It is very French, yet there is a clearly defined Chinese quarter with its own style and culture. The Saigon girls are pencil-thin and very elegant. Americans have moved in here in a big way. I imagine that they are in exactly the same position as the French, and will eventually be kicked out in the same manner.

French is spoken in most places. It is so hot and humid (75F at night and 100F in the day) that I have had almost all of my hair cut off. From mid-day to 5 o'clock the atmospheric conditions are such that most of us drink a lot of liquid, lie down and wait. I was very kindly given a box of my favourite American chocolate bars, but by the time they had been moved from a nearby refrigerator to mine, they had become a runny, brown mass. The chee-cha (gecko) in my room loves biscuit. He lives in a cupboard drawer, and if frightened hides beneath the refrigerator. Street touts are numerous. At the cafés, they can locate Americans in the twinkling of an eye, and plague them unmercifully. The French women are very chic. They do credit to their homeland. The squeeze is on. If a French person leaves the country, they have the greatest difficulty in getting in again. I'm back to the menus of my Paris student days of bread, saucisson sec and red wine. In keeping Vietnam 'in the free world', the Americans pay for the entire Vietnamese army and subsidise all goods that are not luxury goods. With Jeremy I've visited a Cochin China village in the Mekong delta 50 miles away. With less backward cultures in

the countryside here, and with civilised western life in Saigon, this is a very good place to be, and would be a good holiday venue. On seeing me struggling with a pineapple at a table in a small fishing village, a passer-by took my knife and turned the fruit into an object that a French chef would be proud of. He said nothing, smiled and walked away. Mangoes are wonderful, plentiful and cheap. The many boats on the river have woven mat sails held high on the masts to catch the wind above jungle trees. Seafood is sold alive, with crabs trying to escape from their baskets and fish thrashing about in tanks. And there is that catfish of Thailand that looks like a stumpy eel with a moustache. I have visited the

Ecole des Beaux Arts, where, just as in Paris, the students were drawing a nude model as part of a course in French Art. They also study oriental art, and were making a 3'x3' lacquer mural for Brussels. It had taken three months to make and would take another three to fill the inscribed grooves with silver dust and coloured lacquer. The students could choose between Western culture and language or Eastern — not both. The past, freewheeling reputation of Saigon disappeared with the French administration. Matters of a sexual nature are frowned upon. But the phallic nature of past religions is still exhibited in the museum. Jeremy, who, like all the other Americans here, has his apartment found for him, eats American PX food and lives in a very closely knit compound. He was so excited by our little 'escape' jaunt to the delta, that he plans another trip for us into the wilds of the north. I look forward to it as well.

The chee-cha in my room has gone missing. Jim.

CHAPTER 27

Toward North Vietnam and the Gong Makers

APRIL 1959

Permits and passes had been obtained for a trip up-country toward the North Vietnam border. I have never found out exactly what Jeremy does for his country here, but he was very keen to get out of Saigon to see the countryside, the people, and to sound out the feelings that ordinary Vietnamese had for America and the present political situation. So off we set in Jeremy's large American car.

We headed north out of Saigon and then to the east and the South China Sea coast. The houses we passed on the way were made of mud-plastered bamboo with roofs of concertina-folded palm leaves.

Our first objective was the town of Pan Thiet on the coast. A fishy smell appeared in the car. This became stronger, and stronger still, before becoming almost unbearable in its intensity and unpleasantness. The town was famous for its Nouc Mam, a fish sauce. The smell came from wooden vats filled with small fish, water and salt. After fermenting for a year in the open the resultant liquid was filtered, put into bottles or locally made earthenware jars looking like ginger jars, to be sold throughout the region.

In the evening, all available male citizens were being taught to become soldiers. They marched back and forth with some venom to broadcast military music. During the breaks there were propaganda speeches.

We moved on northwards, and were stopped at all bridges and some checkpoint fortifications by particularly ferocious-looking individuals with guns very much at the ready. No-one to whom we spoke in cafés, garages and hotels had a single good word to say about their present American masters. It did seem crazy to see some Vietnamese being trained to fight other Vietnamese who were living nearby on the other side of a line of latitude. Jeremy, who had seen his countrymen as saviours rather than oppressors, was very surprised.

Whereas my Thailand travels were provisioned by the simple fare that we, or the porters, carried on our backs, here I was in a large car containing an ice box full of saucisson sec, French bread, butter, Beaujolais, beer, chocolate, tins of meat and fruit, and other goodies. And in the extreme heat of the day within our metal box of a car, the cold beers were a godsend. Hooray for civilisation, in this case.

The countryside became wilder, with scrub land leading to high jungle as we progressed northwards. A large troupe of grey monkeys crossed the road in front of the car, causing us to halt abruptly. Last across the road was a maimed old male, protecting himself as best he could from the physical abuse meted out by his companions.

Where there was water there was rice cultivation. The whole year cycle was in operation at the same time — soil preparation, seed growing, planting out, green growth, ripened stems, harvest and husking.

Beaches of sand were barren of people. Even at the town of Nha Trang, where there was more marching, martial music and shouted propaganda, and where there was a crescent of fine sand backed by palm trees, only a few children played on the shore. Sun, sand and sea were to be avoided in this part of the world.

Seawater was clear, fish and shellfish plentiful. The fishermen's district had just been burnt to the ground, a common occurrence when the weather was hot and dry and bamboo easily burnt. The residents were living either in tents or under sheets of corrugated iron. The water wagon had just appeared and there was a mad scramble for supplies, with every conceivable shape of container in use. A tapioca vendor watched with amusement.

With French spoken, even in outlying districts, Jeremy

became a Frenchman of convenience. And in a café, where the beer was cheap and the waitress pretty, Jeremy ordered another bottle if his glass was part empty and the waitress stroked him. Stroking was a very successful sales technique. Fortunately there were no other cars on the road when we set off once more.

From Nha Trang we headed inland toward the Laotian border on roads where petrol stations could be 50 to 100 miles apart. Our mountain road sometimes became a spanking new American highway, as one was being built from south to north for military communications. The contrast between new road building machinery and the occasional, ancient, native truck, broken down with its engine spread out on the road for repair, was the extreme one of master race and native.

As map-reader and lookout, I spied an interesting-looking village on the hills a distance from our road. So we stopped to investigate. After a walk through jungle we came to 40' long houses on stilts, made of bamboo and woven straw.

The bare-topped women inside were weaving material, using dyed yarn from Saigon. The pattern they wove was the same as they wore, though theirs was machine made. Weaving was accomplished by attaching one end of cloth to the wall and the other held in tension by the weaver, sitting on a log and bracing herself away from the wall with outstretched legs with her feet in footholds. There was no attempt at speed, though one mother was saving time by breast-feeding as she wove.

My trick of offering sweets worked its usual ice-breaking wonders, and I was able to draw. But immediately Jeremy produced a camera, breasts were covered and the atmosphere fouled.

As we were unable to find rooms in Ban Methout in the high, central mountainous region, Jeremy sought help from those in the local American military mission. These soldiers were living in ex-Emperor Bau Dai's hunting lodge. This consisted of three large wooden buildings on stilts, connected by bridges. All bedrooms contained bath, bidet and lavatory, and all rooms were conveniently connected to their neighbouring ones. As the Emperor's palace was only a short distance away from his hunting lodge, I imagined that his prey would have been somewhat easier to catch inside this complex than outside in the jungle. There were also kitchens, a dining room and cinema.

The amiable coloured Colonel in command, greeted us with the words: 'It's pretty tough in the jungle.' He then opened one of six giant refrigerators to extract tins of cold beer.

His soldiers were given a film show on five nights of the week, ate and drank well, and had access to most modern conveniences.

The speciality of Ban Methout was gongs. Several shops sold them, and they ranged from about 1' to 5' in diameter. Viewed from the non-striking side, they were in the shape of shallow, vertical rimmed plates with a concavity in the middle.

When I was in a gong shop, two mountain men came in to buy a gong to go with the one they had brought with them. The new one had to be tuned, which was done by the shopkeeper. He laid the new gong on a pad, striking it deftly with a metal-headed stick. Once in a while the two gongs would be played together. The rhythm and tunes were most pleasant, and the variety of sounds quite varied.

Gongs were played at weddings and funerals and any other occasion possible. A street vendor even had his own gong musician playing to attract custom. So there was lovely music in the air most of the time.

Beside gong making, timber work and weaving were the main occupations. The only tilled ground I saw was for banana trees.

Many men carried crossbows, and betel nut chewing was much in evidence. I tried a chew of the nut and lime wrapped in a betel leaf, but spat it out very quickly when the lime started to burn into my mouth. The taste reminded me of root beer.

Back with the military people, who had kindly offered us

accommodation, we were fed excellent food, given plenty of drink and shown a rather poor film. And in the morning they gave us sandwiches for our journey and replenished our icebox with beer. We had not expected to be treated with such lavishness in this part of the world. It was interesting to note that although the Americans there were teaching English, not one of them spoke Vietnamese.

We chose to travel now by minor roads, or rather red, dusty tracks. There were not many people around, but we saw two bare-topped women pounding rice, and met a hunter who showed me his crossbow that fired very sharp, flat-sided, bamboo arrows with banana leaf flights.

Houses now were built on the ground rather than on the more usual stilts. This may have been for strength, as in one village, two elephants with their minders were setting out for work in the jungle.

Clothes were simple in the extreme. The women wore only a black skirt, sometimes decorated with yellow and white bands, and the men a codpiece-cum-jockstrap with a little flap in front of it. When, at the side of the road, we saw two handsome women and their men, all smoking Sherlock Holmes pipes in the shade of a tree, it was an excuse for us, too, to take a break and swap pleasantries. Later, we passed an old lady herding cattle. Her dress consisted of a small black turban, a necklace of tiger's teeth, and just a tiny flap, held jauntily from the waist by a piece of string.

Our track led us to a river and a ferry. The craft was on the other side. After some blowing of the horn, ten natives, dressed only in jock straps, and all having long, distended pierced ear-lobes, sometimes with silver rings in them, turned up to ferry us across. They deck-walked the boat by pulling on a wire connected to both banks. A dugout canoe, with poles for propulsion, was also used to aid us in crossing the brown, fast-moving water.

Roads flowed over a landscape of scrub, grassy hills, and through tunnels

of deep jungle. On a rough stretch of track, when a shower of rain turned the surface into a slippery quagmire, the heavy, automatic geared car slid and wallowed as it chose in the red mud.

The noise of our approaching car must have frightened away most of the local wildlife. A large, black monkey did bound across the track in front of us, looking neither to left nor right, and a black bear on all fours didn't quite know where to go in his state of confusion as we approached. Eventually he picked the densest bit of roadside jungle, and crashed into that.

We passed two settlements. One was for refugees from North Vietnam, thriving with the help of modern, American agricultural machinery, and the other for hill people. The latter was half empty, hill people preferring not to be fenced in.

On the first part of our journey back to Saigon, a 90 mile stretch from Ban Mathout to Djiring, we had met only three vehicles, one supplying food to the settlements, and the others transporting workers.

We were told on our return that we had passed through an area of 'rebel' fighting. And we had thought the gunfire was only marksmen trying shoot animals for the pot. Jim.

On my return to the room in Saigon, the biscuit-eating chee-cha had not deserted it. Quite to the contrary, he had been courting in my absence. He now shared his drawer with a pale-coloured mate.

My next mode of transport, the Messagerie Maritime's, white-painted, passenger/cargo ship, the Cambodge, had docked. And as nice as it was to be in sweltering Saigon, the thought of cool breezes at sea and new ports of call beckoned.

CHAPTER 28

By Boat to Hong Kong via Manila

In England, I had been advised not to travel steerage with this particular shipping line. But the MV Cambodge was in, and going to sail to where I wanted to go on the 17th of April 1959. Not only that, it was a cheap way to travel, and now I could at last obtain a rest from some fairly intensive drawing. Moreover, it would be a great pleasure not to have to fend for myself daily in a foreign land, to deal with foreign currency or to struggle with a foreign language.

I somehow thought that steerage was a way to travel that had disappeared some while ago. But this was it, and called 'between decks'. I shared cabin 301 with five others. It is true that six of us could sleep there at one time, but only five could stand up in the cabin at the same time, and there was only enough air in it for five. It was hot in there — very hot. The cabin was just above the waterline at the bow end.

My fellow passengers were a Frenchman, a Chinese, a Sikh, a Japanese and a sort of Indian. I deposited my valuables with the purser.

The vessel was one of several, built to convey the colonising French to and from the Orient. Now that colonists were fewer, and this form of transport less in demand, the shipping line was concentrating on the holiday trade, especially for that between Singapore and Japan.

Our dining room was slightly less cramped than the cabin. And as our first menu consisted of cheese fingers, cabbage and beetroot salad, artichoke vinaigrette, veal stew with fried potatoes, and then pineapple, the omens were good.

But that was not all. A half bottle of red wine was thrown in with each meal. Moreover, as a Muslim jeweller from Ceylon, sitting next to me, announced that he was not allowed to partake of alcohol in public, there was the happy prospect of a whole bottle with each meal. And when I happened to spill some of this pleasant liquid over myself, the smell in the cabin became a little less offensive.

The only bit of deck that steerage passengers were allowed to use was a shadeless cargo hatch. But it did not take long to discover that by going through the galley, there was shade from awnings, breezes and life to be found in the tourist section.

Back in steerage, a notice was pinned up to announce a forthcoming lifeboat drill. With it were directions on how to reach lifeboat No.3, and a chance of life should disaster occur. So, thinking I might find a nice spot to pass the day elsewhere, I took the route somewhat early — only to find a locked bulkhead door barring the way. This was opened for the lifeboat drill, and closed immediately afterwards. So heaven knows what would have happened in a real emergency.

With a lot of Chinese aboard, Mah Jong was played, at tremendous speed, and for high stakes. After four rounds of a women's game, all the contestants stood up, said 'God Save the Queen', and sat down to continue. 'What patriotism,' I said to a companion. 'It was not patriotism,' he replied. 'It was to air the seat.' Then a lady dropped one of the bricks, and picked it up like lightning — in her toes.

After docking in Manila, the steerage passengers were told that we were not allowed ashore. Our kind were liable to jump ship in the Philippines. After kicking up a fuss, I was allowed ashore and took a stroll around the town, only to see that American life and ways had taken over. There were signs of 'Drugstore' and 'Popsicles Sold Here'. It could have been any town in the USA. It was a pleasure to return to the French way of life, cramped as it was.

The stretch to Hong Kong was less pleasant. A large German had taken the place of one of our number in the cabin, reducing the available space. A storm blew up and all ventilating portholes were closed. The temperature rose even further. Tourist Class passengers created a great fuss with their complaints, having paid far more than those in steerage below decks. The quality of the food deteriorated considerably, and an engine failure delayed the boat.

So when we were greeted with a 21gun salute when entering Hong Kong harbour, every one of us, I'm sure, was glad the voyage was over. The salute? It happened to be the Queen's birthday.

CHAPTER 29

Hong Kong

21 April 1959

Blessed are the customs officers in Hong Kong. This is about the first country I have entered where I was not treated as a criminal.

When in Kowloon, the commercial part of Hong Kong on the mainland, I did my usual trick of taking a taxi to size up the hotel situation. I visited a dozen or so, some clean, some dirty, some mean, one grand and several too full of girls. I have settled on the cleanest, the cheapest and the one with the most spectacular view — all rolled together. It is a 1920s to '30s built hotel and has, I am now told, the highest rate in Kowloon for killings and suicides.

My room's period window (masses of glass panes set in a metal frame), covering the entire wall, overlooks the harbour, Victoria, and Hong Kong Island. At night the aspect is one resembling a still photograph of a firework display, it is so bright and colourful. Today, the Peak, islands, and scores of freighters surrounded by bustling junks with wet sails (the monsoon season has just started) are set in a misty stillness — just like a Chinese watercolour.

The hotel has a manager, or clerk, on each floor. He sits behind a desk and is responsible for a 24 hour watch. As in most other hotels in the East, this one is no exception, being a place of assignation. Two of the resident girls have left, leaving the Mammasan with only one at her disposal. The desk clerk takes a commission on any sexual arrangements conducted on his floor. So he is keen for business. Outsiders come and book a room, and order a girl through the clerk. Then, through the channel of the Mammasan, one would be selected by age — not looks, or shape, or colour, or size of breasts, or bottom, or anything else as one might imagine, but age only. When talking with the floor clerk, most of the girls I saw coming in for business were very plain. Only when a middle-aged merchant seaman appeared on my floor and ordered an older woman, did I see a presentable-looking one.

A half-drugged girl, of reasonable looks, patrols my floor occasionally and tries to enter my room by some ruse or other. She would make a good model, but I fear she might collapse and stay if I let her in.

No English at all is spoken in this sector, but I get by as usual. In my favourite café/restaurant, a large galleried place that buzzes with noisy activity, I order a beer and wait. Then, around the tables come the cookhouse boys bringing delightful dishes and shouting out their names. Very popular are simply boiled fish and shellfish with various sauces in which to dip the morsels. One of the most popular sauces is Worcestershire sauce. The Cantonese waiters all want to learn English (I am the only European seen there), so we swap a word or two — usually to much laughter.

New sights, sounds and smells pile one upon another. A woman might eat with a child at the breast. In each of the many newspapers on sale there is a photograph or woodcut of a nude girl — some ravishingly done. People wear pyjamas in the daytime, which makes sense, as the material is so light and bright. Children from a year old are given coins to buy sweets or whatever, so that they might learn the value of money early in life. Mah Jong parlours are everywhere. The clattering noise of the bricks can be heard from some distance away. The Cheong Sam dress, worn by many women, shows about as much of the outside of the leg as possible. With its high, stiff collar, the outfit is extremely flattering, when still or on the move — providing the owner has good legs. But there are also dresses of every other kind to be seen, from traditional ancient Chinese to modern European. Which is lovely for drawing or café-watching.

I have taken a bus trip to the New Territories that lie between Kowloon and the barbed wire patrolled border with China. There was so much to see on the trip. There were

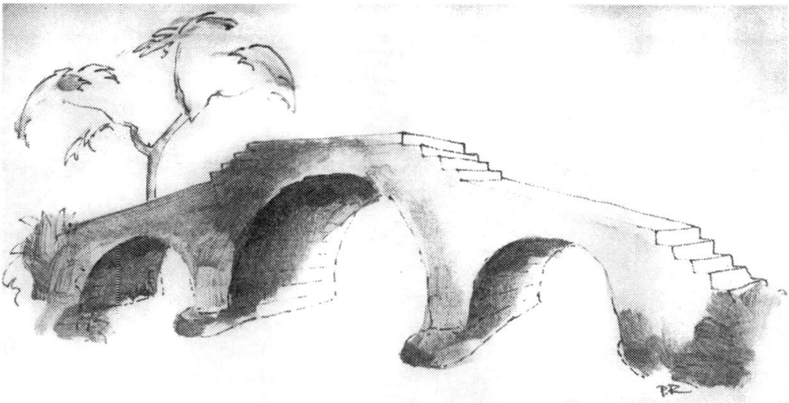

lavatories perched above pools of manure, onions cultivated for their stems only, huge flocks of white ducks and a most elegant, and presumably ancient stone bridge, now isolated from water. It was a three-arched footbridge with steps. 'Hell' bank notes are sold with the denomination of 1000000 (40 for 1d) to be burned with a corpse to ensure an adequate supply of money in heaven.

On Saturdays and Sundays a ferry goes to Tai O, on the island of Tai Yu Shan. I took it. The island is leased from China, and is comparatively uninhabited. It is larger than Hong Kong Island. Being navigated there by a Chinese gave me the greatest feeling of confidence. When a thick fog descended, almost as soon as we set off, our Captain guided us with only a very basic compass and his knowledge of rocks and currents. I had no idea that we were passing through a narrow strait between two islands, and through a sea of swirling whirlpools, green water and translucent jellyfish. As I peered down from the rail at a spot of water in the mist, a large ocean fish rose into the air and fell back with a splash — just in front of me.

We stopped offshore at Tung Chung Wan, where a sampan brought out two passengers and took off three.

After two and a half hours we tied up to the jetty at Tai O. I walked inland along an estuary where houses were built on stilts over water. Some of these houses were real boats, turned upside-down. Most houses were roofed over with arched sheets of corrugated iron. There were no streets or vehicles, only a path. But there was a fire engine (a two-wheeled barrow, painted red) with two men on duty. Both were asleep. And there was a town ferry to cross a few yards of river. This boat was rowed by one woman at slack tide and by two when it raced. Of our passengers, two, who were traders in dried fish and groceries, came ashore to do business.

There was a funeral in progress outside one of the boathouses, where evil spirits were being frightened away in the vulnerable period between death and body incineration. So a band, or sometimes a lone musician, would keep up an incessant noise, and firework squibs cracked with tremendous bangs as they jumped hither and thither to keep the demons at bay, and thus from entering the body.

Water was available from a reservoir in the mountains behind the town. Sampans sailed up the river to obtain supplies from it. The liquid was either carried in tins or jars, but sometimes by just filling the boat. A great deal could be carried in this way.

I walked inland toward the mountains and paddi fields, stopping only when offered tea by a family sitting by the side of the track. But so poor were these hospitable people that 'tea' was only plain boiled water.

Passing over two antique bridges, I saw a navigator who had rowed and poled his craft up the river, to then climb into the water to gather stones from the riverbed.

Farther into the mountains, the steep path was paved with the same river boulders. As the path climbed past paddi fields and waterfalls, it became narrower and narrower. The calm of the surrounding countryside was broken only by the sound of gravediggers in their struggle to penetrate the rocky soil.

So beautiful and peaceful was this most natural and spectacular of landscapes, that I spent the day wandering, sitting, drawing, eating, drinking, and absorbing the very Chinese atmosphere of the island.

Then it was time to return, to be astounded at the skill shown earlier by our Captain on the outward trip, to eat pork and mango ice cream in Victoria, and crabs and prawns with fresh ginger sauce in my favourite restaurant near to the hotel.

I shielded myself from rain and sun with an umbrella. But whenever the sun shone, even the venetian blinds of my room were useless to keep the sunshine out. When a great liner, called the Himalaya, moored outside, blocking the view, the sun still managed to rise above it to scorch the room. Siesta time was the hottest time.

I returned to a tailor's shop one evening to collect the new items he had made for me by taking measurements from the threadbare ones given to him earlier. Now the bolts of material lying on the shelves in daytime had been replaced by human beings - all asleep.

A fellow diner at my café is an artist called Choi Tai Hor, who paints wallpaper for the American market. His work is done on paper, then stuck to silk and rolled on a wooden stick. The rolls he painted, each with seven fish and bamboos, measured 38"x10'. His place of work and living quarters measured 8'x6'.

The locals are very attached to caged birds. I met a lowly sparrow that sat on a perch in the open with a piece of string tied around his middle. He could fly about 18" as if in complete freedom. Then 'bang', he would come to the end of his tether and have to fly back again. He was untied for the night and put into a black bag suspended by string. There were two small holes in the bag for ventilation.

Beggars in this cut-and-thrust society have a poor time of it. One evening in my room, I was watching the passing world and the lights beginning to flicker on the Peak in Hong Kong Island across the water, when a blind man appeared below. He fumbled his precarious way along a wall until he could go no further. He then sat down on the pavement in a corner with his possessions by his side. When the pavement slabs became too hard for his slight frame to bear, he took out a begging bowl and, clearly in pain, slowly put it out in front of him. Then he moved a parcel to beneath his head. Hundreds of Chinese passed by and not one put even the meanest donation into his bowl. So I got dressed, went down and dropped in a few coins. He was now huddled up in the corner wiping moisture from his sightless eyes. He made a noise, possibly of thanks. He was one of thousands of beggars who sleep thus on the streets. Yet despite such aspects of poverty, the overall feeling here is of prosperity, hard work and play.

On a more cheerful and amusing note, I was walking through the New Territories to draw and have tea with the Bishop of Hong Kong, with whom I had travelled as far as Penang. He had suggested that I might like to take tea with him and his wife on any Saturday, should I be in Hong Kong. A car passed by and then stopped. The driver had seen a European walking alone in the wilds and had stopped to ask if I wanted a lift. Inside the car sat my cousin Elizabeth and her military husband. They had no idea that I was abroad, let alone in Hong Kong. We had all been invited to tea.

There were twenty of us there to partake of tea from a

large teapot, and to eat jam sandwiches, scones, butter and cakes. The Bishop entertained thus on every Saturday, inviting the more important members of his flock. The Chinese were not at all sure how to tackle the food on offer.

Lychees are in season, the best, I am told, come from South China. They are bought attached to their sticks by the weight of one catti only, because the Chinese think that more would be bad for the blood pressure, or something. Simple, hand-held, fulcrum scales are used to measure the weight.

The water all around is covered with junks, and I have had a chance to inspect them. They are certainly vessels of great majesty and beauty, and clearly fine sailing ships. There are Hong Kong junks, junks from Red China, junks that are floating shops, hospital clinic junks and junks towed in convoy by very small tugs. There are sailing junks with whole sails and others with tattered sails, often with huge holes in them. Either whole or not, all seem to cleave through the water with tremendous ease. In fact, junks come in every shape and size, and are thick on the water. They move so effortlessly and silently in the wind, or cluster around freighters anchored in the harbour. Each vessel is crowded out with tackle, families and barking dogs. There is washing hanging out to dry, and every other domestic occupation in progress. Home life and business life overlap. So they are not only fine mercantile vessels, but excellent homes as well. There are so many that just why they never collide when under sail is a minor miracle.

I have booked on to a merchant ship going to Japan. But before I leave, there will be time to acquire a visa and cross the Pearl River for a visit to the Portuguese enclave of Macao. Jim.

CHAPTER 30

Macao

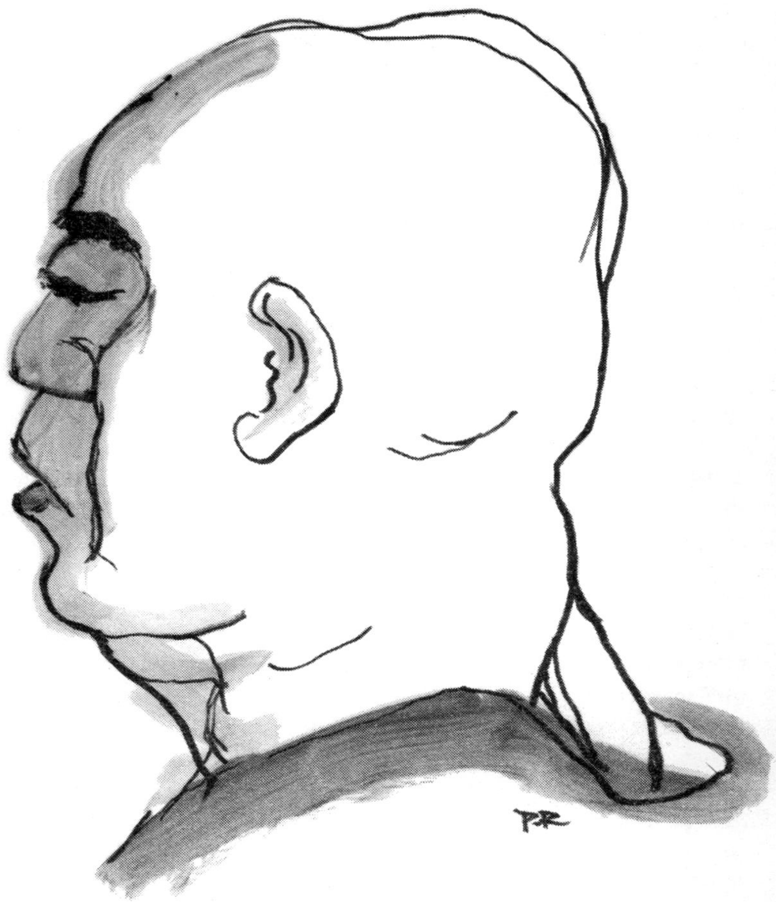

4 May 1959

As the ferry left a Portuguese island on its last leg toward Macao, I noticed a rather nice-looking dog in a basketwork cage on deck. I enquired of my Portuguese/Chinaman neighbour why it was that such a docile and friendly-looking dog should be so caged. 'Oh, he's off to be eaten', said he, and then continued to relate some of the many delicious ways in which to cook dog.

Aboard the ferry was an American soldier on leave. He tagged on to me and would not be shaken off. But it turned out that later on he would be a welcome support.

The delightful, old and crumbly town of Macao, sits on a promontory with the Communist Chinese border close by. So keen were the colonists to make it look like home that the buildings could have come straight from Lisbon. Its closeness to China is very evident. There is a bus service to Canton, yet access from Hong Kong is difficult, with a visa necessary. All the land visible around the town is Communist Chinese, as is the armed gunboat, positioned just offshore.

The Chinese I met in Hong Kong call themselves Chinese, many not knowing that they were residing in a British colony. Those living in Macao call themselves Portuguese, with sympathies lying very much with Communist China.

I suggested to the American that we ate in a restaurant famous for its snake dishes and white pigeon's breasts. We ate the pigeon, but I imagine that the snake could have been the better bet. As we drank plenty of excellent Portuguese red wine, to be followed by strong rice wine from China, I thought that my pen was well geared for drawing. Our rickshaw man was waiting for us outside the restaurant, and suggested that we might want to witness an 'exhibition'- girl-girl, girl-boy. The price was modest and the show would last for half an hour. Fine. Off we went, with me showing our driver my vicious-looking knife, just in case of trouble.

We were taken to a mean-looking hotel near to the shore and to a room with a canopied bed, separated from two seats by a curtain. It was a theatre on a very small and intimate scale.

As my soldier companion was showing signs of having had a little too much to drink, I was to deal with negotiations. Should I pay before or after? Up-front money was demanded, a third of which was to go to the hotel and the rickshaw man, the rest to the performers.

The management retired discreetly and the actors appeared, taking the form of two women, one very pregnant. They disrobed and produced a length of red rubber tubing, on the ends of which had been glued two baby-bottle teats. With great energy and with many cries of 'Hy-ya', they proceeded, in tandem, to insert this gadget into every available orifice.

I regret that I could not take all this too seriously. With the American shouting: 'Get it in baby. Go on, give in. Let yourself go,' and 'Go, baby, go.' I was laughing, and the hand in charge of my pen was a little shaky.

Then the American wanted to join in the act, and a Chinese minder was called in to pull him off. After that bit of excitement, a male Chinese performer was roped in (had he come off the street?) and performed with the non-pregnant girl. After his orgasm, the girl sat on a rather pretty blue pot. I had done 25 drawings and wondered if they would ever manage to hide away in the next batch to be returned to England by post.

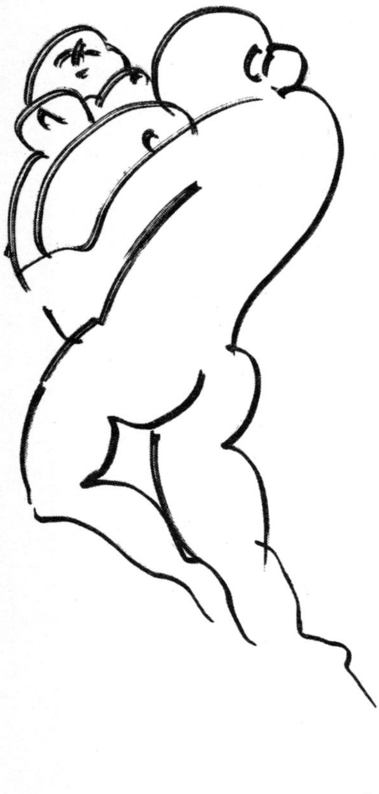

The service man then left Macao, thanking me 'for the best time in my life'.

My hotel doubled as house of liaison and gambling. The walls of the bedrooms had wooden walls in which there were many holes from which neighbours could see and be seen.

The gambling part was most interesting. There were two forms. One relied upon large holes cut through the ceiling into the second floor above. Around these holes were railings, so that the gamblers standing by them could lower their wagers in baskets down to the first floor on string, and haul up the winnings without falling through in their excitement. Police roamed about to see that all was in order. Above the second floor was a restaurant and above that a night club with girls on tap.

One of the forms of gambling involved dice and a green cloth, patterned in domino divisions. I did not understand it. The other, called Fan Tan, was so simple that even I could master it. A square on a table was divided diagonally into four divisions, numbered one to four. The croupier gathered up some ordinary buttons and placed them in a bowl. After bets on the number of one's choice, from one to four, were placed by hand or lowered in a basket from above, the bowl was turned upside down and the buttons left on the table. Then, with a stick, the buttons were taken away four at a time. The number of buttons remaining represented the winning number. The atmosphere and setting of this gambling 'den' would have made an ideal subject for Rowlandson with his pen and brush in 18th century England.

I must return to Hong Kong to board the ship bound for Japan. Jim.

CHAPTER 31

Getting a Feel for Japan

The Havana Maru freighter was lying offshore in Hong Kong harbour. I hired a launch to take me out to her. It was the 10th of May 1959. I was on my way to Japan.

Surrounding the ship were the usual crowd of junks of every size and condition from which cargo was being winched aboard. Then away went the junks, our hatches were battened down, up came the anchor, and we were off.

The ship was Japanese owned, and built by the Japanese for Japanese-size sailors. I used the wash basin when on my knees, and in the shower room the spray head pointed to just above my waist.

I was the only passenger, and ate alone. My personal steward served me what the cook deemed to be European food. All notices were in English, which seemed a bit strange for a Japanese ship

where so few spoke the language. 'It is the custom,' I was told.

The ship was modern and fast. No-one steered. Details of its proposed course were dialled into a mechanism.

With my map of Japan spread out, I was advised where I might visit. I brought my invaluable 'wants' list up to date with phonetic Japanese, and I did some drawings to keep my hand in.

Everyone on board practised their golf swing by hitting balls attached to fishing line out toward the sea. American Ladies' clubs were used.

Iwojima was to the right and Formosa (Taiwan) to the left. We ploughed on relentlessly.

Moji, on the northern shore of Japan's southernmost island of Kyusu, was to be our first port of call. A foretaste of witnessing the country's delicately manicured landscape, appeared in the form of little volcanic islands, each with its bit of clinging vegetation, surmounted by a pine tree or two. Any one could have formed the background of a late 18th century Japanese woodblock print. After that, the town of Moji came as a great disappointment. It was covered in smoke and industrial haze.

An evil looking bunch of dockers came aboard wearing cloven-hoofed rubber boots with the big toe separated from the rest. I was taken ashore to a seedy, run-down city, in an old, coal-burning tug. Since entering the Suez Canal, this was the first time that I had no need to wear a sun hat. Through the dirty air of this antiquated place came the smell of spring flowers. I sensed that Japan would be a country of contrasts. Even the girls were pretty and the men plain.

A cable car left from the top of a building to a 'vantage point'. I bought a return ticket and squeezed myself into a tiny bubble of tin and glass, and was shoved off into space. Up and over valleys and hill crests it climbed. I glanced at the ticket in my hand. It was soaked through with the sweat of fear.

When I returned to Moji, the town was en fête. There were parades, musicians, acrobats and troupes of dancing geisha. About one in eight women wore a kimono.

I was in a coal-burning, island country. And, despite not knowing a word of the language, and also being unable to read anything, I felt very much at home. This place would suit me. I knew it from the start.

CHAPTER 32

Kobe and a Secret Restaurant

12 May 1959

Before disembarking in Kobe, on the main Japanese island of Honshu, a reporter came aboard, mistakenly believing that I was someone of importance. I don't think he understood much of what I said, but he was very surprised that I was unmarried. Most men in Japan are married by the age of 25, and all by 30, he told me.

Ashore, I was met by my mentor in Japan, who had booked me into the smartest hotel — a kind gesture but one soon to be righted. Naoto Izuka is a very nice man, and the President of an importing company. He offered to give me dinner at a sukyaki restaurant.

At 6.30 sharp he collected me at the hotel and we drove through the brightly-lit streets of the city, ending in a dark and dreary quarter. Here he dismissed the taxi and we continued on foot, through muddy ground where there was no pathway, over railway lines and sleepers, and under railway arches.

Beyond the sidings the district became even murkier. It was bleak, dingy and inhospitable. Between shacks lay a dark and narrow passageway, down which we walked. Naoto tapped on an unmarked door, set in a crumbling wall.

The door opened with a creaking sound, and into view came an elegant little courtyard of fountains and ancient storm lanterns. Then, from sliding doors, beautiful ladies, attired in sumptuous kimonos, came to greet us.

We took off our muddy shoes and were taken into a matted and simply-furnished room which, in turn, led to an internal and more lovely courtyard garden of paving and pools.

We sat opposite one another on cushions at a table. Then windows were opened that led to yet another internal garden, so that we could hear delightful samisen music, mixed with laughter of an inebriated kind.

Our food was prepared at the table by the mistress cook on one side, accompanied by her assistant and an entertainer on the other.

Hot sake was served throughout the meal by a girl who entered and left the room on her knees. There was a ritual to this drinking, as we were required to serve her in return. The meal ended with a dish of wild strawberries served with whipped egg white.

After this occasion of delight, we visited a very uninspiring strip show.

My feeling that Japan would be a land of contrasts is being confirmed.

Today I managed to find a Post Office to buy colourful stamps for this unsealed letter. Having decorated the envelope with a fine collection, and to the correct value, I asked the counter girl if she would look for some past issues for me. This she did, and produced three more stamps. I lay out a handful of change on the counter, but she declined to take a single coin. 'No. I make present,' she said. The stamps are enclosed. Jim.

CHAPTER 33

Kyoto and Ryoanji

That first night, spent in a European style hotel on a bed, was to be my last in Japan. Through the kindness and knowledge of my host, I was now to be transported by rail to exquisite accommodation in Osaka, and very much in the Japanese style.

Set on a hillside overlooking the ancient capital of Japan, this inn consisted of a central wooden building with several small pavilions set in its gardens. These gardens were beautifully landscaped, planted with deciduous and pine trees, and decorated with rocks, moving water, pools of carp, and stone storm lanterns. There were arches to purify the mind and body with upturned cones of salt on each side, and straw hanging from above. I settled on a little pavilion house beneath a persimmon tree.

My host ordered lunch, which was put on to the only item of furniture in the room, a large, low, lacquer table. There were two kinds of raw fish, several varieties of seaweed, and bean curd served in several ways, some in a soupy mix. I was told that, for a European, the flavours to be gained from these delicacies, would be almost too fine for me to appreciate. But anyhow, it was the look of the dishes that really mattered.

I was praised for eating up all the fish and for my dexterity with chopsticks. But then I was sure that such praise was part of good manners.

After lunch I was taken to see some Osaka sights. There was a massive concrete replica of a castle, huge wooden gates, temples and a moss garden. This garden was set beneath trees and in boggy land through which flowed a stream. Moss was grown in many colours and states of compactness. All looked as if nature had designed it. It was a garden for the specialist, but charming in its conception. Then I was shown a quite ordinary-looking lake. What was so special about it? Its importance was only as a site from which to view the autumn moon.

After the moon lake, I was taken to a garden that must be one of the most famous in Japan, if not the world, at Ryoanji. Expecting to be unable to appreciate its finer points and significance, I was quite seduced by its amazing simplicity and surrounding aura of tranquillity.

The garden is known for its fifteen stones set in pure white sand. That I already knew. And I had seen the blue stones in white sand from photographs. Well, the sand was not sand. The rocks were not blue. The rocks were the mellow grey colour of rocky Japanese islands (said to represent Buddhist temples), two of which had fringes of moss on top of them. And they were set in off-white gravel that undulated slightly and was raked, possibly to represent ripples, waves or plough. And the garden was far smaller than expected from having seen photographs of it previously.

The arrangement of the 'islands' was in clusters of 5, 2, 3, 2 and 3. They appeared, at first glance, to have been placed almost haphazardly in gravel, which was surrounded on three sides by low, yellowish walls, capped with tiles. On the fourth side, people either stood or sat to view it from a low, wooden platform, backed by priests' rooms.

I was not alone in contemplating this extraordinarily skilful arrangement of shapes and colour. Fellow tourists clustered around, clicking cameras, shoving and squeezing, pushing and chattering. In a few minutes I was quite unaware of my fellow travellers, and completely overcome by the peace and stillness of the scene that lay before me. I felt relaxed in mind and body. To have attempted to capture these feelings in a drawing would have been impossible.

CHAPTER 34

Kyoto, Bath and Entertainment

For a foreigner, the bath must be one of the most unique and fascinating aspects of Japanese life.

My first real experience of a Japanese bath was in the traditional inn in Kyoto. I dressed in a cotton kimono, called a yukata, and walked across the garden to the main building where the bath was situated. I had instructions to sit on a stool outside the bath and to wet my body, soap it, and rinse off with hot water scooped from the bath in a bucket. I was not to enter the bath until absolutely clean. Then I was to climb into the bath for a good soak.

The bath was a traditional one in this most traditional of inns. It was made of pine, and was formed in the shape of a tall box. Beneath the boiling water inside was a seat. The lining of the bath was slimy.

Scorching water was quite normal, and when, at a later date, there was a choice of bath temperatures, even the coolest I found to be unbearably hot. The reason for this heat soon became apparent. Having subjected one's body to high temperatures, the heat became locked inside when thick and well-insulated clothing was worn afterwards. With Japanese houses being unheated, and the paper walls draughty, this was a survival measure. And, in the cool spring weather, I found these very hot baths to be of tremendous help in keeping me warm through the nights.

If water in the baths was far hotter than expected, the towels appeared to be far too small to be of any use. But because the body was so hot after a bath, the moisture remaining on its lobster-red surface seemed to evaporate very quickly. So only the smallest of flannel/towels was necessary.

Throughout my stay in Japan, the form taken by baths and

their imaginative and artistic settings, became a constant delight. They were among the most fascinating features to be seen and used in town and country.

The peaceful atmosphere of the inn was disturbed once each evening when a busload of tourists came to watch a first class geisha dance, sing and play the samisen (a three-stringed guitar). The geisha was dressed in most elaborate clothes of startling colours. She wore a magnificent coiffure arrangement of black hair and combs, and her face was painted white with scarlet lips. She walked with little steps on high blocks of wood. In fact, she made a fascinating sight, but hardly a human one. I was told that with new labour laws forbidding young girls to perform, and train in manners, music, dance and decorum, the number of first class geishas was in decline.

I went to the geisha theatre at Gion. But before the performance there was a tea ceremony, conducted by a geisha, so that the audience could relax, purify and prepare for the

performance to come. The geisha conducted much ritualistic handling of pots, water, cups, stirrers and green tea powder. After this, everyone sipped tea very daintily. Then, as the show was about to begin, all tranquillity was forgotten as the audience jumped up and stampeded toward their seats, pushing and shoving as they went.

The auditorium was square, with the audience sitting in the middle or on a balcony. On either side of the stage were geisha singers/chanters. Actors reached the stage through the main body of the theatre on raised projections. As costumes dazzled the eye, and acts of dancing, music and song took place, scenery was moved at lightning speed on lifts and tracks.

Back at the inn, Naoto and I dined on raw fish, soy soup, eels grilled in their own oil, lobster in shell and salmon in skin - all delicious. Later, my bedding was rolled out on to the matting floor of my room and, as a matter of courtesy, I was offered a girl for the night. She was not to my liking. It was somewhat of a task to decline in what I hoped was a well-mannered, and roundabout Japanese way. It must have been accepted, as later I was invited to the staff's quarters to watch a boxing match on television. I think I was considered to be somewhat of an oddity.

CHAPTER 35

Amanohashidata and a Look through the Legs

May 1959

It is wonderful to be looked after by a mentor. I get to see aspects of life and places that only a local knows about. And, almost more important, I learn about customs and what to do and what not to do, which, in a country like Japan that is so different from any other, is vitally important. But it is wonderful to be free, to go at my own speed, be able to stop and draw whenever I want, and to be beholden to no-one. I am now free again. And I can leave the town for the country with a map in my hand.

One of the three most beautiful sights of Japan is Amanohashidata. Those aboard the Havana Maru ship almost insisted that I go there. So I took a three hour train journey from Kyoto toward the north-east to reach it. I wondered, as the train reached the Sea of Japan, with Korea and Russia beyond, just how much prettier it could possibly be than the sights from the train of spectacular rock cliffs, with pine trees and sea breaking over smooth rocks.

Amanohashidata translates as 'ladder to Heaven'. This ladder takes the form of an isthmus of land and bridges, running diagonally across a bay. When seen between the legs from a nearby hill it represents the escape route to heaven from earth. So naturally up I climbed and looked. And as much as I could stretch my imagination I could see no more than an attractive isthmus of land in the bay. But later on I might mischievously have gained an interpretation when seeing a postcard of a geisha with feet apart, having raised her skirts for the view.

I have now had a good look at the countryside from the

train, and have absorbed a wealth of first impressions.

We passed through a landscape of hills covered by forest. The timber was harvested in poodle-fashion, with trees cut in fields, as though they were grain.

A cow pulling a one-furrow plough through a small rice field, seemed to be a satisfactory and economical way of tilling and fertilising such a small area. Other crops on low ground were soy and cereals.

The Japanese are an industrious race, yet they are late to rise in the morning. I have seen little work done before 8.30 o'clock.

The houses of the country people, who work on the flat land between the many steep hills, are made of wood, wattle and mud, topped by deep thatch. Some have either tile, or tile combined with thatch roofs. There are also some modern coverings of corrugated asbestos. Where framework is

covered by paper, it is protected from the outside elements by overhanging roofs or glass. The textures, inside and out, are varied by combining different wooden surfaces, woven bamboo, glass and yellow cement. These surfaces might be broken up with balconies and railings. Though many rooms must be very draughty, the station waiting rooms are stiflingly hot and airless.

Fire is clearly a great hazard with houses and whole districts constructed of inflammable material. So, for quick detection, the many fire stations incorporate lookout towers in their design.

Raw fish, raw and almost raw eggs, feature on most menus, with rice eaten after everything else. The texture of the rice lies between the separate grain variety of China and the glutinous kind of northern Thailand. It is easily grasped by the wooden, 7" chopsticks, presented in a paper packet, to be used and then discarded. Though less easy for me to use than the 12" washable kind of China, they do grip the food more easily.

The weather here in late spring/early summer is cold, windy and damp. The roads are full of holes and covered with mud. So what with the damp, dirt and chilly winds, the boiling hot evening baths and subsequent body cocooning, make life bearable. I feel the cold, especially after months of tropical heat. Jim.

CHAPTER 36

Matsue

21 May 1959

Precision timing on the Japanese railway system is quite amazing. Arriving early at the station for the train, as I invariably do, I was the only one to be seen around when the train was due in ten minutes. So I thought that my timetable information was incorrect. Five minutes to go and there was still no-one to be seen. Three minutes to go and some passengers appeared. Two minutes, and some more arrived. Then, on time to the second, the train arrived at the platform. The Stationmaster checked his watch carefully. We departed on the very second designated. So when the Post Office at Amanohashidata had no stamps, and I was told that my letters would go down the line by rail to a post office that stocked them, I knew that their transportation to England would be handled with efficiency and precision.

Station names are displayed in Japanese and English, which is making rail travel much easier for me. On enquiring why English should be used, I was told that it was 'to please our former guests'. There are no porters at the stations, as no-one seems to carry much on trains. What they do carry is put into a square of cloth and tied with two knots.

I now had another chance to concentrate on countryside and customs, despite the conductor sometimes standing up to give a long talk.

Orchards made a wonderful sight, with each individual fruit held in a paper bag. And to protect the fruit from marauding birds, ugly-faced kites were held loosely on string between poles. Very cleverly, the kites rose slowly in the breeze and then dived.

Where road surfaces are dry, pedestrians wear gauze masks to filter out the dust. And there are many more kimonos to be seen in the country than in town. But whereas so much of the countryside is a pleasure to look at, on wooded hills are sometimes to be seen large and brash advertisements on hoardings. Once again, it is that strange Japanese combination of extremes — art and vulgarity.

Vulgarity and art is once again exemplified by the approach and then entry to my present inn in Matsue. From a dirty and unprepossessing back street I entered a building of orderliness, calm, cleanliness and efficiency. Even toothpaste and brush are provided, but in a land of paper, there is none in the lavatory.

Not a splash of paint has been used in my charming room. Yellow cement and wooden walls divide rooms. Paper-covered sliding screens divide mine from an outside balcony. The wall that separates the room from the passageway is made of board and sliding screens, all surrounded by a black lacquer strip.

My room furniture consists of a wall cabinet and shelf, a raised wooden platform, specifically for a vase holding a small tree-branch and a piece of sculpture, a low table, a screen on which to hang clothes, a basket for shirts and socks, a telephone, and a lady's dressing table (looking more suitable for a doll's house) with its mirror covered by an embroidered cloth.

And there is the ubiquitous calligraphic painting on the wall (two in this instance). To my western eye these works of art look rather alike. But to the Japanese they are very important objects of veneration. Why? Well, they are generally executed by famous religious men. It is not necessarily the 'words' that count. The dexterity, the calligrapher's reputation, the work's inner meaning, the time of day or night when brush was put to paper, the surroundings and the state of the artist's mind, all count toward the fullest appreciation. Most of the works consisting of deft black ink swishes on paper, backed by a silk mount, are stamped with one or more signature seals, stamped in an almost luminous, orangey/red colour. It is a colour that, to my mind, adds greatly to the interest of the black and white work.

At night time, the table is put aside, the bed arranged, and a paper-covered electric light connected. This lamp has a switch to select florescent for general light and reading, and a low power bulb as a nightlight.

There are seven, finely woven, black canvas-bound floor mats — the number seven being something to do with the tea ceremony. Beds, taking the form of two well-stuffed eiderdowns covered with cotton sheet are laid on it. To cover the bed are two more like them. The pillows are stuffed with rice husks, which might as well be gravel.

Outside shoes are left at the door of the house, to be replaced by house slippers. These, in turn, are left at the room door. Only bare feet or socks touch the mats. In the WC, soft slippers are exchanged for wooden ones.

I am no longer surprised by Japanese food. Breakfast here was cockle soup, dried seaweed, and spinach in cold milk. And

I have come to the conclusion that as my natural metabolism will have difficulty in gaining enough strength from such fare, I must eat up everything that is offered. I still have great difficulty in consuming raw eggs. But I have to do it.

I have had a good look at Matsue. The town of two-storied buildings is built between the sea and a large lake, typically dotted with islands of rock and pine trees, reminiscent of the garden at Ryoanji. In fact, almost everywhere I look in Japan, I see the typical landscape of pines, rocks, water, woods, steep hills, wooden houses and rice fields. I feel at home in it after looking at so many late 18th and early 19th century Japanese woodblock prints.

Since leaving Osaka, I have now seen a little of the Sea of Japan coastline, experienced life at inns and enjoyed rail travel. I have just been looking at the map, and decided that my next move will be to take the railway south over the mountains to the Pacific side of the country, and visit Hiroshima and Myajima. I know that the train will leave exactly on time. Jim.

CHAPTER 37

Myajima

The train had to cross the spine of mountains that runs through the central part of the main island of Japan — Honshu. As the train climbed into cooler air there were still signs of the famous cherry blossom of Japan. So, at the end of May, I did manage to get a glimpse of it. But, honestly, it was exactly the same as any other cherry blossom. Lovelier by far was the wild wisteria. It covered buildings, grew over shrubbery, stood as a tree on its own and, in one instance, in full bloom, covered the entire top of a tall pine tree

The train had to climb and descend some very steep gradients. So, as a safety precaution, or as a way of saving the brakes perhaps, the train would rush helter-skelter down a hill and then up a steep branch line on the other side. Then it would roll backwards, until regaining the main track once more. These switchback tactics made good sense, and were a lot of fun.

Most of the mountainsides were covered with forest. But the rivers, though picturesque, did not compare with the rushing and brooding waters of the jungles of northern Thailand.

I did not really want to witness the sadness of Hiroshima, where the atomic bomb had been dropped fourteen years earlier, so changed trains to go directly to Myajima.

Myajima Island was another of the three great sights of Japan. And being of great educational importance to the Japanese, the place was inundated with uniformed and regimented parties of schoolchildren. So I based myself on the mainland and took a ferry to the island to appreciate its natural beauty.

It began to rain, so I bought myself a cheap and beautiful umbrella, made of bamboo and varnished paper. Even when whole, about 1/10th of the rain that fell from above carried on to me below. Then, after a clash with another umbrella, and a brush with an overhead branch, about half of the rain fell through.

So into the rubbish bin it went, after a very short but stylish life.

I found shelter from the rain in a café, where the fried oysters were superb. As I ate them I noticed outside that the rain had forced many flies toward the seawater. Then down came a host of swallows to eat them. Other than swallows, the only birds I had seen during my whole stay in Japan were two dark pheasants.

Tourist spots around the world tend to be spoiled by their very popularity. Myajima was no exception. Although one could appreciate great arches standing in the water, the site was cluttered with souvenir shops. Even the ornamental pools were coated with empty tins and floating orange peel. So I left the crowds and walked inland to hills of magnificent pine forests and aromatic undergrowth. There I was completely alone. Thinking that the island really did offer real peace and beauty after all, some music reached my ears from above. It was the voice of Frank Sinatra, coming from a loudspeaker at the top of one of the trees. I decided that places of natural beauty in Japan were not for me.

My next general target was to be the southern island of Kyushu. Intending to reach there by train, I discovered that a ferry was leaving Myajima for the Kyushu town of Beppu almost right away. It was still raining when I bought a second class ticket and boarded a particularly ugly boat.

CHAPTER 38

Ibusuki and the Volcano

MAY 1959

As I boarded the vessel on my way from Myajima to Beppu, I, and all the other second class passengers, deposited our shoes in an anti-room, or rather anti-cabin, before being led to a very much larger cabin, where the matted floor was covered by sitting or sleeping people. I managed to find a bare patch and lay down on it with the blanket and rice-husk pillow provided. It was very cramped, and those squeezed tightly around me were either sleeping and/or belching, snoring or farting. The atmosphere was rather thick, and none too pleasant. But that is the way people travel by ferry here.

We docked in Beppu at five in the morning. The town is on the east coast of Kyushu Island and famous for its hot springs and pretty women. I may return.

After a quick look at some steaming, sulphurous water bubbling from the earth, and a visit to a public lavatory where the hole in the floor led directly to a running mountain stream beneath, I boarded a train toward the southernmost tip of the Japanese archipelago. My plan was to map-read with care throughout the journey, and return by train to what I thought might be a nice fishing village in which to stay. I did see one, but its name is not on the map.

However, it lies between two marked villages. So I have a good chance of getting back there.

I knew that there was a live volcano close to the town of Ibusuki. But it was either too cloudy to see on arrival, or perhaps it was shrouded in its own smoke. Anyhow, after looking at two unsuitable inns, I have settled on a brand new Japanese-style hotel, as an experiment.

Rooms here come in many price ranges, and as far as I can see, they are all exactly the same. Only the aspect matters. I have chosen the cheapest and most remote as it is a pretty heartless place anyway. The raw fish served is excellent, the service abominable. But the nicest part is the bath. It is more like a swimming pool, in looks and size. I wallow around in it on my own, getting redder and redder. I am the only customer in this large hotel.

The sand on the shore outside the hotel is black and volcanic. From it rises steam and sometimes scalding water. In a hole in the sands that she had scooped out, an old lady was taking her 'cure'. She was dressed in a black cotton yukata, and wore a towel headdress. Shielding herself from the cloud-covered upper elements with an umbrella, she was painstakingly using a wooden spade to fill up her own 'grave' with hot, black sand.

Further along the shore, fishermen were offloading smooth shelled cockles from their boats. Nearby, many kinds of fish, fish heads and bones were being dried in the hazy sunshine. All were covered with flies. I was offered a half-dried cockle — which tasted like a half-dried cockle. Inland, women were tilling the soil.

It is easy for me to eat in the town as the café dishes are reproduced in plastic and displayed in the window with the price attached. All I have to do is take the waitress outside and point. My evening meal ended with what looked like ice-cream, but it was the ubiquitous raw egg, this time beaten with crushed ice.

As I found the town not be of great interest, I decided to return by train to the fishing village seen from the train coming south, and was on my way by foot to the station, when I saw the volcano — far, far closer than I had imagined. It is a multi-peaked volcano and exuberantly active, yet there was a party of schoolgirls, crouched at its very foot, drawing pictures of it on large sheets of paper. So keeping a wary eye on the volcano's belching peaks, I bought my ticket, and was very glad when the train pulled out of range. Jim.

CHAPTER 39

Minacannes

MAY 1959

I had not much faith in my map when map-reading on the train south to Ibusuki. But the village I liked the look of lay next to the marked Takanabe Station. Those in the Ibusuki hotel who gave me advice, thought that another station, called Kawanabe, would be a better one at which to alight for where I wanted to go. This request caused much head-scratching behind the ticket counter at the Station. Anyhow, with the ticket in my hand, off went the slow, stopping train. Incidentally, stopping trains are by far the cheapest. The more circuitous the route, the cheaper they become. Second class on faster trains is more expensive, more comfortable, and offers booked seats, loudspeaker information, and girls serving refreshments.

Most passengers aboard the train were reading magazines, no two of which seemed to be alike. To the western view, they were being read back to front, with the characters running from top to bottom, left to right. There appears to be no set rule about direction of writing and printing. Two directions are often used on the same printed page. I enquired about this, and was told that the Japanese just knew. A lined exercise book is laid on its side and written on from top to bottom left to right down the lines. The recording of dates, such as the franking of a stamp on an envelope, is difficult for me to understand. Sometimes it is written as day, month, year, sometimes the other way around. It can also be written as the number of years that the Emperor has reigned, followed by the month and day (34 5 22). I could not decipher 30V 59 18 24.

I was making a note of these exotic matters in the carriage, when a canary burst into song from beneath its cage cover — much to the amusement of the passengers. It even continued singing when the steam train climbed slowly and laboriously through a tunnel, when its human companions were coughing and spluttering in a carriage full of steam, fumes and smoke.

We reached Takanabe and then stopped at my destination, Kawanabe. No wonder there was head-scratching at the Ibusuki ticket counter. The only building in the entire landscape of rice fields was the station on which I stood. But, of course, there were platform officials. They, at first, suggested that I walk the two miles to my fishing village. Then one thought of a taxi, and went off to find one.

An hour later I was on my way to Minacannes. It is the custom in Japanese villages, that when there is no inn, travellers are put up in a home. I have been given a room above the general store.

No-one speaks a word of English. But with my minimal command of Japanese, I was immediately understood, well cared for, and treated as a friend. Paper walls are usually in poor condition. Paper forming the walls of my room is no exception. So, as I write this, I can see several silhouetted shapes of spectators and many eyes peering through the tears and holes - wondering about this very strange foreigner in their midst. Not all the inhabitants have come to see me. Fishermen are mending nets on the shore. Jim.

CHAPTER 40

Toryama Hama

LATE MAY 1959

Although the station was possibly named Minacannes, I think I am in the village of Toryama Hama. Anyhow, wherever I am, it is exactly the place that I was looking for. Fishing is the main source of income for the village. The fleet consists of several motor and sailing vessels. Next in importance after fishing come seaweed collecting and drying, catching crabs, and harvesting large winkles and oysters from around boulders on the shore. This is done by women who, when bending over to prise, scratch and scrape, are almost indistinguishable from the rocks.

The main variety of fish caught, look like a cross between a mackerel and a whiting. Served cold with apple slices, it is cooked by scoring the skin, rubbing in salt, and then grilling. It is a tremendous relief to be able to eat food that my stomach understands. Other fish caught are blunt-nosed dolphins. These look miserable, but are beautifully coloured with yellow bellies and iridescent blue/green above. The rest are swordfish, ordinary and hammerhead sharks, and odds and ends.

Outside my shop is a square with a fish market at one side. In the afternoon, a man appeared on a bicycle to set up a small stage in the square. On the platform he placed a pack of coloured cards depicting Victorian people in a melodrama setting. Then, to a good audience, and beating and shaking a tambourine for dramatic effect, he told a story, taking away the front picture to reveal the one behind it as he went along. It was an exciting performance for us all. Having told his tale, he began to sell transparent toffee twirled on to sticks. Presumably, the performance was only to attract an audience to help sell toffee. Sadly, only two children bought his toffee,

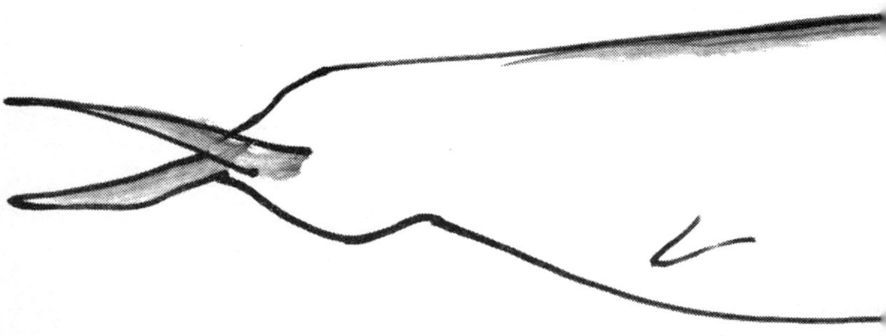

and off he peddled to the next village.

After dark, a cinema screen was set up against my shop. The entire populous turned up with their mats to put on the dry earth in front of it. I was made a guest and not allowed to pay. The show contained a Russian cartoon of ancient origin, and a Japanese blood-and-thunder costume drama, with a pause between each reel. In the costume drama, the hero was thrown off a great precipice half way through and turned up again at the end to slay the villains — to great cheering from us all. There was, however, one most interesting incident. There was a love scene when a young man kissed a girl. The audience, silent just before it, burst out laughing. Finally there was a circus story featuring the Russian, Popov, at the start of his career as clown and slack-wire artist.

This place is worth all the beauty spots in Japan put together.

With my breakfast comes coffee. This says 'dehydrated' on the tin, but could have been made of anything burnt and ground up. It is supplied with hot water. Then there is soup,

made of fish, onion and scrambled egg. And, as always, a raw egg, complete in its shell, but this time accompanied by soy sauce.

My evening meal might be made up of soup with eel, cold grilled fish, pickled turnip, raw fish served with delicious onions, and always rice and tea. An American in England told me to take vitamin tablets when living in the Japanese countryside. I bought some in Thailand and now take them as my meat course. But I am fast losing weight.

The Japanese love wood in every shape, size, colour and smell. They handle it with great love, care, and reverence, not to mention skill and craftsmanship. Well, I was directed to the WC soon after arrival here, to be confronted by a wooden wall of delightful, yet complicated design. I could see no door in the panels. It must have been somewhere, as there was no other exit or passageway. I was getting desperate. It was a Japanese wood puzzle to be solved quickly, which I did, and only just in time, by sliding a horizontal panel batten to the side. The hidden door then opened, to reveal the usual hole in the floor over a tank of 'night soil'. All was well.

The iron bath in my general store is shaped like the deep bowl of an English copper, and heated from below by a wood fire in the same manner. The water prepared for me is, as usual, scorchingly hot. In this case the liquid covers about a third of my body. So, having pushed down a floating disc of wood, to protect my bottom from the hot metal beneath, I eventually emerge two thirds white and one third red.

The village has a form of Post Office. The clerk had to telephone elsewhere to enquire if he was allowed to handle a letter to England. Weights and prices were relayed to him. But when my letter was placed on the weighing machine, it registered nil. For fish — fine. For air mail letters — not so good. Jim.

CHAPTER 41

A Fishing Trip

I had made great friends with the patriarch of the village. And, in the course of a sign language and drawing conversation, I indicated that I would like to accompany the fishermen on one of the boats that I had been helping to haul up and down the beach. So it was settled. I would have to rise at 01.30 in the morning to be ready for a 2 o'clock departure. There were to be three others on board with me.

It was dark. By the light of a hurricane lamp, we loaded baskets of line. Then, with shoulders to the hull, we pushed our boat down the sands and into the sea, clambering aboard as best we could in the blackness of the night.

Then, with poles and oars, we eased the boat toward deep water, bumping and scraping over submerged rocks as we went. This made a lot of nasty noise, but having seen the boats built in the village, without recourse to anything other than board fitting exactly to board, I knew that no damage would be done. Once clear of the rocks, the propeller was lowered and we were off, steering a compass course.

At dawn the engine was cut, and we set about catching the mackerel/whiting live bait. With each of us using a hand line of hooked lures, it took very little time to fill a compartment with thrashing fish.

As we were obtaining this bait, a lovely dark sea bird circled our boat and landed in the water nearby. Showing off like a puppy, it swam towards us in fits and starts with its head beneath the surface, making bow waves as it approached. When it was very close, one of the crew found an oar and hit it on the head, breaking its neck. With the bird still alive, he took it from the water and threw it to me. I tried to show my disgust at his conduct, and offered to put it out of its misery. But the fisherman took the poor creature away and threw it back into the sea. It lay there on its back, flapping its wings, with its broken head and neck dangling below the surface — where it drowned.

I tried to dismiss this unpleasant incident from my mind, but couldn't.

With enough live bait on board, we headed for the fishing

grounds to catch what we could with a long line. The line was attached to an end marker, which was cast overboard. This indicator was a bamboo flagpole, weighted at the bottom, and given buoyancy by a glass ball attached one third of the way up.

Setting a compass course, the main line, from which hung hooked and baited secondary lines, was paid out from one basket to another. Occasionally another flag marker was added to the line. And sometimes a glass ball was attached to keep the line near to the surface.

After four and a half miles of line had been released, the other end flag marker was cast overboard. Now we slowly retraced our course, looking for a bobbing flag or ball to indicate a catch.

The weather began to deteriorate and became unsuitable for catching sharks. But many of the yellow-bellied dolphins had taken our bait.

By the time we had reached our starting flag, the weather had become so bad that the Captain decided to end fishing for the day. Hooks were unbaited and the line curled back into one basket after another.

It was becoming very cold and wet aboard. I noticed that callouses and other bits of previously hardened skin were being torn from the fishermen's fingers and hands as they pulled in the line. They were quite cheerful about it. It was hard to draw, and my paper was getting wet.

With all line, flags and glass balls aboard, we met another fishing boat returning to Toryama Hama. It was not until I was able to see daylight under her hull that I realised how much we, too, were being tossed about on the waves. But having seen the boats made by a craftsman with the skill of a fine furniture maker, and without even having to use paint, pitch or caulking, I had no fear whatsoever for our safety.

We regained port, and I was about to join the crew operating the windlass ashore to haul the boat up the beach, when the Captain, with some ceremony, took me to his house and offered me the use of the family bath. It was a most welcome offer. This bath was another of the 'copper' variety, but a very public one. Not only did his family come to watch me in it, but also passing neighbours would break their journey to come close, peer at me through the window, bow and smile.

My only embarrassment was that the flea, or fleas, that joined me on the ferry trip to Beppu, had by now feasted off me so heartily, that to my audience it must have looked as though I had the measles. Fleas were a natural hazard, with so many hiding places for them in woven mat flooring.

Having restored heat to my cold bones, and when the catch had been taken to market, I was invited upstairs to sup with the family. With the others eating typical Japanese food of fish scraps, pickled vegetable and rice, I was presented with a whole grilled fish — a rare, pink specimen that we had just caught at sea. It was delicious.

CHAPTER 42

Two Visitors and Money Trouble

LATE MAY 1959

I have been constantly surprised by sudden switches from kindness to cruelty, elegance to the tawdry, beauty to ugliness, artistry to crudeness and so on. So I should not have been upset by what happened next.

I had spent the morning drawing shell-gatherers on the shore. Then, having bought and eaten some lunch, I was about to relax until the tide went out once more, when there was a tap on my door.

Two Japanese men in sports clothes had come to see me. This was not at all unusual, as foreigners were completely unknown here, and I had by now become a friendly object of curiosity. I knew that they were amateur photographers, as I had seen them on the shore taking photographs of waves breaking over the harbour protection wall.

We sat on the matting floor of my room, and for nearly an hour conversed with a few words, signs, gestures and drawings. There were, quite naturally, long and boring silences. I thought it might be impolite to conclude matters.

It seemed that they had heard I was an artist and had come to see my drawings. I was unprepared to show unfinished work, and besides, when my drawing paper is handled by others, dirty finger marks appear, as do small folds where thumb presses against finger when the sheets are moved. So I just pretended not to understand.

It was uncomfortable for me to sit cross-legged on matting for so long. I wanted to be away. Then they showed me their identification papers. They were policemen.

We chatted once more. Then they demanded to see my

passport. I showed it to them, and they took copious notes. But they seemed unable even to recognise their own visa. So I pointed it out and suggested other items that might be of interest to their superiors. Now, once again, they demanded to see my drawings. Still furious that they should appear in the first place, disguised as civilians, I refused, and sat tight. Only force would have budged me or made me show my work.

By now we had had quite enough of each other. They rose, smiled, bowed and departed. It was the only time so far that I have had difficulty in reciprocating the formalities of etiquette.

The incident ruined what was left of my day. But I suppose I had experienced another aspect of Japanese life. Did they imagine I was a spy?

I had begun to think of moving on, but realised my cash supply was on the low side. So I made off by foot to the nearest town.

After only an hour or so on a road that passed through rice fields, a cyclist overtook me, halted, and asked, in excellent English, if I was a foreigner. We walked to his home.

It transpired that he taught English at schools over a large area. His children were to follow in his profession. Apparently, my local stationmaster, who had been unable to understand my request for information about banks along the line, had telephoned this teacher to see if he could be of any assistance.

My new companion bemoaned the fact that accents were now American, when, in his youth they were English. He really wanted to hear me speak. So I did.

We sat in his house, with his wife remaining on her knees just outside the room, ready to fill our glasses with lemonade when wanted. It was really nice to be able to speak English once more.

The schoolteacher then took me around his small village to see the houses and rice planting in the fields. We walked past an open window where there were several naked ladies washing and bathing — a sight that I took in my stride.

We said our farewells, and he told me where to find a bank. But on reaching it, they had not seen travellers' cheques before, so could not change them.

Finding just enough money in my pocket for a rail ticket to a town called Miyazaki, I went there, to discover a bank-like building where all the occupants were standing by their desks doing exercises to music. When the music stopped, I was told that it was not a bank at all, but that there was one down the road. This one really was a bank, but they, too, were unable to deal with travellers' cheques. However, the manager escorted me to a bank that did. And all was well at last.

I have paid for my room (an absolute pittance) and tomorrow take the train north to Beppu. It will be a sad occasion to say goodbye to the lovely people of Toryama Hama. I have had a great and interesting time here. Jim.

CHAPTER 43

Beppu, Bath, Volcanic Sand and a Beautiful Girl

Beppu, being famous for its volcanic baths and pretty girls, was clearly a place to be sampled. I arrived there by train from Toryama Hama.

Seeing a lot of naked bodies through a street window, I took a chance and walked into the imposing building. There was a lady at a desk inside, who took my money in exchange for a cake of soap and a hired towel. There were plenty of naked men about the place and some clothed women.

Lockers, with ingenious keys, provided safe storage for my kit. So, with the locker key attached to my wrist by a rubber band, I entered a hangar-size room where there was one bath, the size of a swimming pool, brimming with steaming water, and another, smaller one, containing not quite so hot water. So far so good.

Beyond these was an entrance to a sort of dungeon of black sand. The sand was littered with solitary human heads poking out of it. There were also prostrate bodies being buried or dug up. All were being fussed over by women attendants.

For my bath, the 8"x18" towel was, as usual, to be used first as a flannel and then squeezed out to become the towel. It is surprising how little towelling is necessary to dry the body after a very hot bath. So I took my bath beside the cooler pool by first sitting on a stool outside it, soaping and sloshing off with a small tub of pool water. Then I lowered myself gently into the scorching water. All this was done as per the normal procedure. But within view was the dungeon of sand. It had to be experienced. So, as if I knew what I was doing, got out of the pool and, as nonchalantly as I could manage, strolled into the inner sanctum of steam and black sand.

A woman conducted me to an available area, dug a trench, and indicated that I should climb in. She then placed a towel over the more delicate parts of my body and, after putting a sort of pillow beneath my head, proceeded to cover me from neck to toe with the very hot, black volcanic sand.

Beneath the ground the temperature was extremely hot. But when the surface of black sand above my body was raked by my woman attendant, the heat beneath it rose tremendously.

The temperature now became unbearable, but there was no way my Japanese vocabulary would run to: 'Help! For heaven's sake get me out of here!'

The only way to save myself from being cooked to death was to try and struggle free. But, with the great weight of sand on top of my body, this was almost impossible. I was about to be roasted. However, starting by wriggling my toes and fingers, then feet and hands, followed by arms and legs, I managed to break free — much to the amusement of everyone around.

I had been buried in the pink and exhumed bright red. Even all my fleabites had been burnt off by the heat and gone.

After washing the black sand out of all bodily crevices, I was even able to withstand the temperature of the hottest bath. I was quite proud of that.

Bathing next to me was a Japanese man who spoke English with a strong American accent. He asked if I had enjoyed any of the beautiful girls for which Beppu was famous. He would be my guide. We dined in a small restaurant where the menu was written in French, and then made our way up a hill to an inconspicuous looking inn. I was shown to a small bedroom and told to wait while arrangements were being made. My guide appeared again and we drank beer and talked. Then into the room came the prettiest girl I had seen in Japan. She was tall, slender and quite stunningly lovely.

My companion retired. The girl and I disrobed and entered a private bathroom that was many times larger than the bedroom. The floor and walls were of mosaic, all decorated with woodland scenes in pastel colours.

Sunk into the floor, off-centre in the room, was a small, sunken, oval bath of clear, hot water. The only illumination for the entire room came from coloured lights beneath the water. The setting was one of underplayed, Eastern artistry.

The girl washed and entered the bath. It was one of the most beautiful sights I was to see in Japan.

Later, the girl asked if she could accompany me on my journey through the country. The pimp, who had by now reappeared, demanded to come, too. I had some difficulty in escaping from them both, to be able to travel on my own again. But Beppu had fully lived up to its reputation for baths and beautiful girls.

CHAPTER 44

Shikoku

2ND OF JUNE 1959

There were few passengers aboard the ferry from Beppu to Uwajima, on Shikoku Island, so I was able to lie on a couch at the side of a large cabin. And one or more of my fellow passengers (in the shape of fleas) were left behind as I had hoped when, at dawn, we entered a most impressive, fjord-like harbour, surrounded by ochre coloured, terraced hills. I disembarked.

So mountainous was the hinterland, that not even the Japanese had been able to tunnel through to make a rail connection from Uwajima to Kubakawa, on the Pacific Coast at the far side. A bus service took its place.

As the crow might fly, the distance was only about 50 kilometres. But the bus voyage, often along the course of a rock-strewn river, took 5 1/2 hours. It was slow driving (20 mph) on narrow roads up and down around mountains. But the river we followed must have been teeming with fish. To catch them, some anglers sat in their boats with rod and line, others stood in the water, casting from there, yet others would lie in their boats with head over the stern peering into the water through a glass-bottomed bucket, to spear the fish from there. But luring the fish to within spearing range required the expert use of bait on a line. This line was attached to the big toe. So bucket-holding and spearing was accompanied by much leg-in-the-air jerking.

The villages through which we passed had a Tibetan feel about them. Streets were narrow and the roofs of houses had exaggerated overhanging eaves. Swallows, with an orange band between wing and tail, swooped around the houses.

Our bus was a large one and ill-suited to the narrow roads. With size taking priority, oncoming traffic would usually have to retreat to a passing-place. The conductress would often have to get out and help the driver negotiate tortuous bends. Crossing narrow, wooden bridges was another hazard for the driver, and the incessant rain yet another. But without the rain I might not have seen workers in the fields dressed in straw. They looked like small, mobile haystacks — thatched people.

Shikoku exhibited an air of primitive timelessness. So when the ride came to its zig-zagging end at Kubakawa, I was sorry it was over.

I stayed at a charming inn, and was served a pink-fleshed fish for the evening meal. I imagine it came from the river, perhaps even lured by toe line and speared. A lady of the place came to spend some of the evening with me.

After a short train journey to Kochi, I boarded a ferry to travel back to Osaka, on the main, Honshu Island. This time I had a second class ticket that provided me with a bunk in a four-berth cabin. I had some difficulty in fitting into a bunk that was made for the size of a Japanese person. And when a storm blew up, any form of rest became impossible.

Tea was provided on board, but no food. Fortunately, I had bought some English chocolate and American, packaged raisins. Well, I thought they were, until closer inspection revealed that the Japanese-produced packages were almost exact replicas of the originals. The Japanese makers had altered a word, or even just a letter on the packet to deceive the buyer, though what was inside was just as palatable. Whisky, car tyres, cigarettes, lighters, and in fact many products (except cars) appeared to be true originals, but were not.

I made enquiries about this deception, and was told that before Japan became an industrial nation, anything of worth had been imported from either England or America. So the packaging had been retained as 'a mark of quality'.

The Japanese love photography. And they love noise, too, especially when a microphone can be connected to a loudspeaker. They carry their own music around with them. Noise blares out in trains, buses, in public places, beauty spots and over the countryside. The age of communication (and noise) has arrived here in a big way.

The ferry arrived in Kobe at 06.30, and I sit writing this on a small piece of deck in the sunshine. Next comes a short trip to Osaka by train. Jim.

CHAPTER 45

A Hillside Inn, a Rock Bath and Geishas (2nd class)

My friend and mentor, Naoto, met me at the station and announced that I would not be staying in an Osaka hotel, but accompanying him to a country inn instead.

Once again I was able to travel on the wonderful railway system. Food and drink was served by streams of charming waitresses (incidentally, the brassiere does not seem to be in general use in Japan), warning was given by loudspeaker on the approach to all stations, and the 'WC Engaged' signs were visible from all seats.

We alighted at a small station and set out on foot up a steep pathway. It was heavy going as the rocky track twisted up the mountainside. A fast-flowing stream ran down through a ravine at our side. Then, with no other habitation in sight, we reached the inn, built into the rocky side of the mountain.

Tea was brought to us, and a considerable amount of negotiation took place between my host and the management. Telephone calls were made. And when the matter in hand had been settled, it was time for a bath.

The large bathroom cave had been partly cut out of the mountain stone. The pool/bath was made from rearranged rock and filled with the customary, scorching hot water. The polished rock surround to the bath was punctuated by concavities. These received drips of cold water that fell from fissures in the rock roof above and from stalactites.

When in the hot bath it was possible to position oneself under drips of freezing water, so that the body was hot and the head cold.

Everything must appear natural, even if it is contrived. The

Japanese manipulate nature to suit their artistic whims. Here was an example, and a wondrous one at that. Another was a Japanese garden I saw of pines, rocks and water. When I glanced upwards, there were two gardeners hanging on to a high branch, pruning pine needles with secateurs — just to make nature look more natural.

The negotiations had been about how to obtain two geishas for the night. It had taken time to find two that were available at such short notice, and who were also prepared to climb up the mountain in costume. The prospects were not good.

Having bathed, we dined. Then the geishas entered, dressed,

not in full, first class geisha attire, but less flamboyantly, as befitted their second class status. They had not been trained to wear elaborate robes, to dance, play the samisen or to sing. Their lot was to smile, converse, fill up the sake cups, and to keep their employers happy in other respects. With neither in the first flush of youth or beauty, one was distinctly plainer than the other. I had first choice.

After bathing early in the morning, Naoto and I took the train to Osaka, and I carried on to Tokyo.

CHAPTER 46

Tokyo and Kobe's Ascot Girl

EARLY JUNE 1959

Taxi travel in Tokyo is enormous fun. I have discovered the drivers' aims. To be frightened is to fail. To stop is to lose. To speed is to be safe. To pay attention to verbal or physical abuse from the customer in the back is not to be contemplated. To frighten passengers is success.

City life I find to be tiresome after being part of country life. So I have visited Nikko to see the shrine and the 'See No Evil...' monkeys, and tried to map-read around Tokyo, where streets have no names and houses no numbers. So I have been quick to arrange an onward passage by ship (the Delos), sailing to Australia on the 9th of June 1959.

I boarded the Delos in Yokohama to learn that we would be stopping briefly in Osaka, sail on to Kobe, enter dry dock in Tamano, and then set sail for Hong Kong, Borneo, New Guinea, Brisbane and Sydney.

Although it sounds to be an excellent itinerary, my few fellow passengers are mainly rich Australians with great thirsts. I do not think I will be able to keep up with their consumption of alcohol, or have the wherewithal to pay for it. And there is also an untrained child aboard. I'll see what happens.

We have had some excitement. The rainy season, with its sudden storms, is about to start. Despite this, our pilot declined the services of a tug as we approached the dockside in Osaka. Then we were struck broadside on by a sudden gust of wind, and in the ensuing panic, the tug was engaged and our ship saved. There is a sense of unease aboard.

It was a pleasure to avoid the heavy drinkers on the Delos and to be ashore in an Osaka bar with a motley crowd of sailors and girls. A Greek ship was in, and the sailors, who were mostly related to each other, were having a fine time boozing, laughing and fingering the girls. I, too, had a fine time, sitting at the bar with the charming and very civilised Madam in charge, talking, eating, drinking and drawing. A Greek sailor was most distraught to learn that his girl of a previous visit had moved on to another bar in another town. But he then found another girl, who performed just as well.

Kobe is only an hour's sail away.

There is something very pleasant about this busy port of Kobe. Long and narrow, it nestles between green hills and the sea.

I bought a rather rudimentary map (always the very first thing I do when reaching a new place) and on it saw that the town had four theatres. So off I went to investigate. Three of them turned out to be cinemas, but the fourth was 'live'. Although it was early afternoon on a working day, the place was packed with customers.

The performance consisted mostly of dialogue between pairs of middle aged Japanese men and kimono dressed women. There was laughter, but I understood nothing.

In the interval a juggler appeared to twirl around a couple of wooden sticks. Each was about a foot long and had an orange tip, like a large matchstick. Suddenly the sticks appeared to become glued together, and the act continued with him still juggling but trying to force the sticks apart. The combination of deception, acting and juggling was delightful. Then there was more dialogue, so I left.

Japanese towns are full of narrow alleyways. And down one of these in Kobe was a wall poster depicting two crudely painted nudes. The establishment was constructed of wooden scaffolding, with a black-painted, corrugated iron outer skin and roof.

I opened the door, paid a modest sum, and found myself in a theatre with a wide but narrow stage. There was no curtain, but at the back of the stage stood a painted piece of scenery, depicting a red striped fence, on which had been glued hearts set at different angles. Behind this was a drape of black cloth, on which had been attached silver-painted, astronomical signs.

A raised projection from the stage invaded the audience area of bare earth. On either side of it were four rows of mat-covered benches.

Overhead was a fine ceiling of striped cloth, and at the back of the theatre, high up and through a hole in the corrugated iron, was a single, man-operated spotlight.

The audience was totally male, and some were a bit the worse for drink.

The show was continuous, and when I arrived at two in the afternoon a plain and tuneless songstress held the stage. Her act was followed by another, a female singer, calling herself Tokyo Anna. She, too, had trouble with her voice.

Next on to the boards came the comedian-cum-stage manager, who, in the middle of much to say, had his words drowned out when a storm broke overhead. He carried on manfully, but the rain falling on the

corrugated iron roof won the contest.

He was followed by four acts of girls stripping. The first girl was very pregnant. She discarded a veil and nothing more — which was quite enough. The next was notable only for a pair of enormous eyelashes. The third sported a huge pair of breasts that she swung up and down and around and around with great abandon. The last stole the show. She came on stage wearing an "Ascot" outfit, topped off with a wide, pink, gauze hat. She wasted no time in divesting herself of every item of clothing except for the hat and a pair of high, platform-soled shoes.

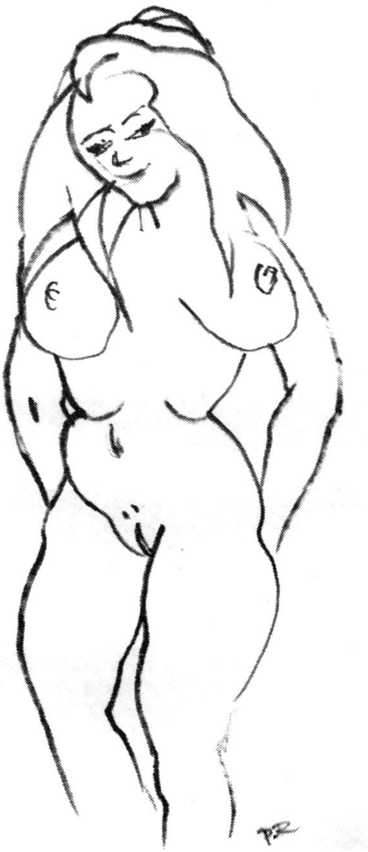

She then proceeded to swagger down the raised gangway among her audience, sneering and snarling at the leering men on either side, who were now rather put on the defensive. She had caught sight of me drawing, out of the way and on the side, but near the front. She sauntered back along the gangway to the stage, turned swiftly, and ran back down the gangway to leap high into the air, to fall in the splits position. The crashing noise was great, and anything could have broken. After gasping with astonishment, there was huge applause from us all.

Then, fixing her eye on me once more, she came over to position herself on stage directly in front of me, crouching with her legs wide apart. She wanted to be recorded for posterity. The audience loved it. My hand was shaking.

Her act was not yet over. Reaching centre stage once more, she plucked out a pubic hair and threw it into the audience as if it was a flower. With so many performances per day, I wondered just how long a career she had in mind. Jim.

CHAPTER 47

Dry Dock and back to Shikoku

When the Delos had sailed on to Tamano and dry dock, the passengers were asked to leave the ship for three days. I decided that the backward island of Shikoku would be worth another visit. I took a ferry there, and a train along the Inland Sea coast, to note the name of any pleasant-looking agricultural and fishing village positioned between mountains and sea. I found one, got out at the next stop and boarded a train back to Iyo Michimasangawa.

The inn I chose was not clean, but the proprietors and atmosphere delightful. I was an object of interest to the children of the neighbourhood, who followed me everywhere in the village and countryside. They were a happy lot, appreciative of my drawings, but a distraction. One of the industries of the region was salt farming. This was done by pumping seawater over bunches of twigs. These took the form of Christmas trees. The sun then evaporated the water, leaving salt adhering to the branches. Other occupations were rice cultivation, drying sardine-like fish in the sun, and papermaking.

A small factory converted old paper into new, and there were many craftsmen making hand-made paper by that so-easy-looking method of dipping a fine sieve into pulpy water and withdrawing it coated with a film of wet paper. The covered sieves were then laid out in the sun to dry. Making paper such as this is one of the great arts of Japan. Nowhere in the land was paper to be seen in dull form. Varieties, textures and colours were legion.

I was writing in my room in the late afternoon, when my bed was laid out on the matting floor and a mosquito net hung around it. The net was thick, and made from foot wide strips of blue gauze, hand-sewn together. I was then treated as one of the family for the rest of the evening.

After breakfast of green tea, a very small quantity of rice, dried seaweed, salami, shredded raw cabbage, cucumber, bean curd, vegetable soup, and a raw egg in a bowl, all served with two kinds of soy sauce, I was off again, back to Tamano by ferry, to see what progress had been made with repairing the Delos.

When ferries leave a pier in Japan, cymbals are clanged to warn of departure. Then Ould Lang Syne is played over the loudspeaker system as the boat leaves the quayside. Coloured streamers are held between passengers and their loved ones left ashore as a last possible contact between them. In this case the ferry trip was a short one. So it was a major and tearful send-off for a very minor journey.

There was time for me to have my hair cut before the Delos was due to leave, which gave me a chance to see how such an operation was conducted.

The work was done entirely by young girls. Each customer chose his treatment from a menu. They all chose a haircut, face shave, shampoo, shaving of eyebrows and forehead, a shave inside the ears and ear holes, and ear de-waxing. As these operations took some time to complete, the customers slept throughout the treatment — in the vibrating chairs provided. With just a haircut demanded, my fair locks were the only pale ones to fall among the black hair on the floor.

I also had my shoes repaired by a cobbler who had made all the shoes for sale in his shop.

The Mitsui shipyard at Tamano had just completed the

construction of a mighty oil tanker for Texaco. From keel laying to completion had taken four months and nine days. Yet I saw no-one working in a hurry or at night.

I had experienced several earth tremors during my stay in Japan. In fact, there was at least one a day somewhere in the Japanese archipelago. A recent one, I was told, killed a few thousand people, but had been hushed up so as not to alarm the tourists. When a 'quake does take place, the major damage is caused by fire, then the gas mains fracture and districts of wooden houses become engulfed in flame. As another earthquake was predicted, I kept a constant eye out for escape routes.

There was some kind of delay in repairing our ship, which gave me a chance to draw in one of Japan's oldest salt fields near Tamano. Here, seawater was splashed over sand in the 'field', and then allowed to evaporate in he sun. The surface, now with a high concentration of salt crystals, was scraped up and placed in troughs. Seawater was then added and the concentrated solution pumped away for the water to be evaporated from the salt.

It was time to leave Japan. The ship had been made ready for sea trials and departure. It was a sad moment for those of our number who had acquired girl friends in the town. With charming and touching sentimentality, the girls came to the dockside to bid tearful farewells with gifts and flowers for their men.

On waking the following morning, I found that we were not on the high seas at all, but anchored off Kobe. Someone had added water to the engine lubricating oil. This was thought to be sabotage. So, with police and insurance people aboard, the polluted oil was laboriously hand pumped into tins and taken ashore. With clean oil aboard, we were off. A steward provided me with a screwdriver and I lengthened my bunk.

In England I had bought a ticket for sea travel to my original target country of Japan. Having reached there at last, and by many other and unexpected means, Japan had amply lived up to my expectations as a place of art and culture, simplicity and charm. I had not expected the extreme and sometimes brutal contrasts. It had provided me with a lot of unexpected surprises, and I had found many and varied subjects to draw. My visit had been a success.

CHAPTER 48

North Borneo

16 June 1959

Bright light, reflected from my drawing paper, has almost blinded me. There have been times when I was quite unable to see. So a period of rest aboard ship has come at just the right time. When we get to Hong Kong I will find someone to look at my eyes.

Because of our delayed departure from Japan, certain agreed ports of call in the South Pacific have been cancelled. With this information, combined with the prospect of being unable to escape the drinking school aboard, I have contacted the shipping agents in Hong Kong, requesting a transfer to another Australia bound ship. Word has come back that this will be possible if there are no further delays at sea and that I act with speed on arrival. I will act like lightning.

I will be escaping from 1. A New Zealander, who still speaks with a Lancashire accent after 25 years in the Antipodes. 2. A retired Australian politician with his wife and daughter. He has a pear-shaped body, talks but doesn't listen. The daughter needs a 4 1/2" breast lift bra and a leg shave. His wife has barely spoken. 3. A late middle-aged widow who has sunburned legs up to where they meet and insists that we all view her suppurating blisters. She peels off skin in public. A piece of it once narrowly missed my glass of beer. 4. A very nice Aussie and wife who come from Sydney and rue the day they brought their perishing two little boys with them. We all agree with their feelings. 5. The Captain, of whom we see little. 6. A relief Captain who will take over in Hong Kong, and is mad on electrical gadgets and plays loud music. We wonder how much of this noise we will be able to stand. Then, of course, there's a tall English artist who is a bit aloof and is inclined to keep to his cabin. He was once seen giving a child a boot up the backside when it was running about and making an unholy noise late at night. The child has since quietened down — especially when he is around. For this action he is regarded by many as a hero.

The cool weather of Japanese early summer has been left behind, to be replaced by warmer air and higher humidity. Now will come the temperatures of the northern tropics, equator, southern tropics, and then the Australian winter — and all in about six weeks. Jim.

26 JUNE 1959

I have changed ships from the Delos to the Aros in Hong Kong. And I am very pleased to have accomplished the changeover, mostly by water, without a hitch. This one is more my kind of boat. It is older, built in England for the Atlantic trade in 1948, salt-encrusted, quieter, but hotter. I have moved from the atmosphere of a Swedish yacht (she was registered in Sweden, but built in Denmark) to that of a working coaster.

I share a cabin with a 40 year old Australian bookmaker and exporter of cocoa. He hails from Rabaul, in New Britain. He is a nice fellow and most informative about the islands. So I learn a lot from him. He kindly leaves me to work in the cabin during the mornings, but pees on the lavatory seat. He disembarks in Rabaul.

I did get the chance to visit an oculist in Hong Kong, and have been given eye ointment and told to rest my eyes by keeping them closed or in the dark. There is time to rest them, as our first stop will be some distance away in North Borneo. There we collect timber from a place called Kennedy Bay. After that the itinerary looks equally interesting.

Kennedy Bay is located on the east coast, and has a village on the shore, backed by dense jungle. The village is made up of a few atap-covered (atap is a kind of rush or palm leaf) shacks on stilts, two houses, one shed and a sawmill. It is an American run business, using Filipino labour (there are no natives of the region in sight), cutting Sorya hardwood from the jungle for export. When the great boles of timber have been cut and trimmed, they are floated on the water of the bay, chained together and, in our case, jockeyed into position alongside the Aros by a little, terrier-like tug. The timber is to be turned into plywood in Australia.

With the logs lifted aboard, we were off through gaps in coral reefs and around densely wooded islands, to anchor off Silam, in another small bay of North Borneo.

Silam, the native name for the local village and nearby mountain, had once been a large and thriving township. But after much pillage from Filipino marauders and pirates, it has been reduced to a total of four households. The pirates, it seems, are disinclined to attack if resistance is expected, but in this case, the whole town had been taken over by pirates quite recently, who slaughtered a European and many natives. Those who escaped fled into the jungle behind, where there are elephants, orang-utans, and giant leeches.

The logging station is not unlike that at Kennedy Bay, but here a muddy camp has just been created, and there are bulldozers, trucks and a crane visible. They also have a lathe, as, in such an isolated location, replacements for broken machinery have to be made on the spot. The logging operation is run by a charming couple. He is English and she a chief's daughter. They work for a Singapore-based American company, who lease tracts of jungle from the British, Queen's Colonial Government, who tell them the amount, size and type of timber they can extract. Royalties and taxes are levied at every turn. Even a postal package, collected from a Post Office a day's boat run along the coast, is released only after a tax has been paid. Despite all the pettiness and avarice shown by the government, and the ill-feeling that it engenders, the white people love it here.

The logs, weighing from three to six tons, are aboard, but unfortunately the ship's carpenter has been slightly crushed when loading one of them. So we have moved, and anchored again. He has been taken by boat to the far side of an island for observation at the 'hospital'.

There has been comedy as well as tragedy. The Captain ordered the lowering of a lifeboat so that he could check on the carpenter's condition. The boat's engine was started, but stopped when touching the water. When going once more, it, with the Captain and its crew, chugged out of sight around one side of the island. Almost as soon as they were out of view, the mended carpenter and his crew appeared in their boat from the other side of the island. It was never explained just why the Captain's lot returned all of six hours later.

It is the 4th of July, and with our cargo of logs on board we are off again, this time to Medang, on the north coast of New Guinea.

We plunge through the tropic seas, where the flying fish glide and frigate birds and brown boobies sail on the wind. And my eyes are better. Jim.

CHAPTER 49

Madang, New Guinea

The Aros chugged its way through the Celebes Sea toward Medang, on the Australian Trust Territory of Papua and New Guinea.

A party was held when we crossed the equator, for which I made a four-headed mask. Much aquavit was consumed and many glasses lost when the spirit was downed in one and the glasses cast over the shoulder and overboard.

Shortly after crossing the equator, we passed Manum volcano, which rose alone from the sea, roughly in the shape of a cone. This was a volcano as it should be, with graceful lines sweeping up to a belching apex. From its peak rose a column of white smoke, or steam, that formed its own cumulonimbus, anvil-shaped cloud in the sky above. This was nature, seen not only at its most elegant and colourful, but most awe-inspiring and powerful. Having been very close to an active volcano in the Japanese island of Kyushu, and experienced many minor tremors in that country, I was apprehensive of beasts like Manum, and glad when the ship had left it well behind.

Madang, a small town set in lush jungle, with roads surfaced with sharp, crushed coral, held a small population of friendly natives. Always more than willing to pose for an artist, it was some time before I realised that many of my male subjects were women, and women, men. The facial characteristics of these Melanesian people

were very alike. The distinctions became clearer when away from town. Here, women, free from the missionary's gaze and obligatory dirty clothes, allowed their pendulous breasts to hang, as nature intended. They carried their babies or heavy loads in nets on their backs, slung from a band around their foreheads.

Their long thin cigarettes were rolled in newspaper. This was a commodity so much in demand for the purpose, that the cost to them was as much as a shilling a sheet. A gardener, who worked for an Australian, was more than pleased to accept newspaper instead of money for his labour.

As I strolled around to see what might be drawn, I was greeted everywhere by a cheery 'Hello Master'. And the children would follow behind and present me with hibiscus flowers.

Industries were the production of coffee, cocoa and copra. But because climatic conditions varied from tropical heat to cool mountain air, almost anything wanted could be grown.

Copra, the dried kernel of the coconut, was gathered from where the ripe nuts fell to the ground. Husking them was carried out on the spot by the deft twisting of the nut over an iron spike. The flesh was then extracted from the nut and taken to a drying shed. Here, the halves were laid on mesh racks over a fire, fuelled by the shells. The dried, half kernels were then taken away to have coconut oil extracted from them.

The natives were small, and varied in colour from tan to black. They lived in simple village houses made from split bamboo and leaves (limbom and sack-sack). Inside each were a few pots, bowls and ritual objects.

Once, while making a drawing of a woman and child, her man appeared straight out of the jungle next to us, wearing nothing but hibiscus flowers in his hair and an animal's tooth through his nose. He carried a beautifully-made bow and a quiver of arrows. Some of the arrows had three pointed heads for fish, others barbed points for pig. The bowstring, made of creeper, had a flat forward face. The arrow flights were also made of flattened creeper. He cut a very fine figure — exotic, naked, yet quite natural.

The spoken and written English was pidgin. A notice, written as SIMOK ITAMBU, translated to SMOKY TABU, or NO SMOKING.

Here we offloaded the bulk of our cargo. There were large, missionary-boiling cauldrons, Land Rovers, tinned fish and diesel engines. All were for the Lutheran Mission Limited.

How, I enquired, did a mission, who's job was to convert the heathen to Jesus, and was presumably funded by charity to do so, have so much money with which to buy these expensive goods. I was told that they were Limited Company Missions, and that they ran large plantations with cheap, 'converted' labour. So it was straightforward business, conducted with the help of God. There was also a Society of the Sacred Heart of Jesus Limited.

With many islands within the administrative area, an airstrip had been constructed among these simple people to cater for 80 flights a week. With so few roads, airways link the towns.

Having taken on cargo of coffee from the highlands and cocoa from the coastal region, our next port of call was to be Lae (pronounced Lay) to the south of Madang. But as several ships were waiting offshore to load and unload, we headed instead for Rabaul, in New Britain, another Australian Trust Territory, linked with the Papua and New Guinea administration.

CHAPTER 50

Rabaul, New Britain

2 July 1959

Rabaul is a larger town than Medang. It lies beside a volcano crater bay and, instead of the bamboo and leaf buildings of Medang, corrugated iron is used instead. The people in a native village I visited are just as friendly, and look different from the Melanesians seen before, only when they have bleached their hair to a ginger colour.

There is plenty of volcanic activity about. Large eruptions keep happening, when a piece of land is likely to sink into or rise from the bay. Only a mile or two away from the town, sulphurous fumes emerge from a crater. When this volcano last erupted, it deposited a thick layer of pumice dust over the entire community. A resident volcanologist takes the temperature of these dormant beasts, hoping to predict eruptions. He is particularly interested in an inactive cone, which he believes will be the next to burst forth.

I love trying to decipher pidgin English. On the roadside was a sign: HAUS I TAMBU LONG OL BOI. I now know that tambu is tabu. Housey tabu along all boy, means building prohibited to all — or Keep Out.

My cabin companion, who has reached his destination and disembarked, took me around to meet his mates. I did not know that he was an inveterate Pommie-hater. So his friends were very surprised to find that he had made friends with one. But even though all seemed to dislike the English, they would not allow criticism of the royal family. I'm beginning to learn something about Australia and the Australians. The combination of them hating Pommies, and me liking them, should be an interesting one.

My roommate's replacement has come aboard. He is a tractor driver from Tasmania. His intelligence is limited to remembering how many bottles of beer he drank in Sydney, how many in Adelaide, and how many in Brisbane, etc.

Of my other fellow passengers I have little to say. There is a New Zealand couple, an Oxford-accented Australian barrister and his wife, who must have been glamorous in her day, a nice Australian couple of the same middle-aged vintage, a plump lady from Sydney who is friendly with the crew, a dear old lady sheep farmer, an old man we seldom see, a couple who run an export business in New Britain, and the young Swedish Captain and his wife. The Captain is very keen on sea manners and creating traditions. He and his wife do not expect anyone to start eating until they do. They can keep us waiting for up to a quarter of an hour in the foyer until they arrive to lead us into the dining room. When they kept us waiting last evening, I led everyone in and we sat down. The Captain and his wife may, or may not, have received the message. (They did.)

Word has come through that congestion off Lae has eased. So we are off again, back to New Guinea. Jim.

CHAPTER 51

Lae, New Guinea, and Honiara in Guadalcanal, British Solomon Islands

Lae was so spread out that I had to walk two and a half miles to reach the Post Office. With a larger town went less happiness. The inhabitants were a rougher bunch than the Melanesians I had seen before. Newspaper-wrapped cigarettes were just as popular as in Madang, but here they were carried stuck into crinkly hair, like large hatpins.

Europeans kept to their club swimming pools, sensibly not venturing into the muddy estuary or clear seawater, for fear of crocodiles, sharks and deadly water snakes.

I missed the charm of the simpler natives in Rabaul and Madang, and was quite pleased to be on my way again.

We now sailed through the Soloman Sea toward Honiara, on the island of Guadalcanal in the British Solomon Islands. Although not the original capital and administrative centre, the Americans had created the town as a wartime base, and because of its greater amenities, was now the new capital.

With the Aros anchored offshore, I took the stevedores' boat ashore to find friendly people once again. The architecture was more pleasing and, most noticeably, there were birds, masses of them, and singing.

An immigration officer, who came aboard to check our papers, told me of a village some two miles out of town that would provide me with subjects to draw. As soon as I set off on foot, a most cheery bunch of natives became my 'guides' — even though there were no turnings off the road. On my left was dense jungle, full of squawking parakeets and cockatoos, and on the other, dead coral, sand, and a clear ocean teeming with fish and dangerous sharks. The sand was littered with sharp pieces of coral of every

shape and size. The road was surrounded by tropical beauty, created by a temperature of 80 degrees, in 80 degrees humidity, and with 80 inches of rainfall. The immigration officer had spoken of two miles. So after walking for about four miles in the heat and humidity, and still seeing no sign of habitation, I turned back toward the ship.

The Solomon Islander's reputation has always been one of indulging in the odd bit of savagery. But I found the people to be gentle, and was told that even in the jungle there were no dangerous animals.

As I was contemplating this ideal state of affairs, a speeding ambulance forced me to jump to the jungle verge. It was rushing toward the town — trailing blood from beneath its rear doors. Then, from the direction of the town, came policemen in various forms of transport.

I learned later that a man was leaving church and had his belly slit open with a copra knife by an unknown assailant. The criminal was apprehended.

Isolation was such that a letterbox, nailed to a palm tree, stated that collection would be at 2am on Tuesdays only.

Australian accents mixed with English in the town. The inhabitants had run out of meat, and the meat ship was not due for several days. But they had beer, and were proud of their 'record' consumption of it, which, in the scorching and humid weather was quite understandable. Even the natives were complaining of the heat.

As Guadalcanal had been a major area of conflict during the Second World War, when the Americans and Japanese used it as a military base and battleground, the jungle was littered with shell cases and the wrecks of aeroplanes, tanks, tractors and trucks. I saw many of these. So the main export from the island was scrap metal, and would be for many years to come. Other exportable items were the more natural ones of shells from the sea, and copra.

It was time to move off again, with the next port of call Vanikoro, in the Santa Cruz Islands.

CHAPTER 52

Vanikoro, Santa Cruz Islands

21 July 1959

Two new passengers came aboard at Honiara, on Guadalcanal. It pleases one, an Australian, to run down England and the English. I know what to expect when I get to his country. The other is a missionary, a real one. I asked the missionary if there were any missions on the island of Vanikoro, where we were about to visit, and was told, no. 'There is no copra on Vanikoro so they can't go into business there. They only 'mission' where they can make money.'

Having now seen so many natives hiding their nakedness in filthy clothes at the bidding of the church, and learned of the Missions' grasping ways, my feelings about the Christian religion are hardening by the minute.

Because of a storm, we won't be in Vanikoro until tomorrow, 22nd of July.

Vanikoro, an island measuring 5x12 miles, is all that a tropical island of the South Seas should be. It is lush, mountainous, densely wooded, and with coconut palms growing on the sandy shoreline where coconuts have been washed up by the sea and taken root. All this is encircled by a coral reef, half a mile out to sea. The reef protects a lagoon of brilliantly clear seawater in which live countless brightly-coloured fish. Flying foxes take wing in the evenings, and money birds (known as such because the natives once used their scarlet plumage as currency), fly around like kingfishers. The sky is almost cloudless. The air is warm and moist, so that when it rises under convection to the mountain tops, clouds form and rain is deposited on the highlands at the rate of over 360 inches a year.

I hear that the mail leaves Vanikoro once every nine weeks — with luck. So I won't be posting this letter from here.

The only way to reach Vanikoro, part of the Solomon Islands Administration, is to sail through a very narrow opening in the surrounding coral reef. With this accomplished, we have anchored off a settlement that consists of a workshop, freezer, two houses for Europeans and some native huts. There is also an 'office'. This is the Customs House, Immigration Office, Post Office and headquarters of a logging company.

Two Europeans live in the only settlement. One of them, a Polish/American, sold me stamps from the Post Office part.

In the logging camps of the interior there also live several other souls. My cabin mate disembarked to join them. The steward was delighted to be rid of him. Not only did he get drunk every night, but was covered with bleeding spots, not to mention that he spat blood from a nasty-sounding cough. I, too, was glad to see him go. But in one respect he was useful. When drunk, he was wont to beat his fists and head on the table and exclaim: 'That's as bad as back of Bourke! That's as bad as back of Bourke!' When sober, he told me that Bourke was in the outback of Australia and was about as ' back of beyond' as one could get. So I have decided to head for Bourke as soon as the ship docks in Brisbane.

The smoking tastes of Vanikoro natives differ from those of their fellow Melanesians in New Guinea. Instead of newspaper, here they favour the leaves of lined exercise books as wrappers for their cigarette tobacco. The width of three ruled lines is considered ideal for a cigarette. Women favour pipes.

Is this really the paradise island it appears to be? Well, it is very isolated for a start. There is much drunkenness and fighting. Mosquitoes abound at low altitudes, and malaria is rife.

The timber workers are able to escape on the boat that calls every nine weeks, but most choose to complete the 22 months of their fares paid, engagement contract. Some stay for longer.

It was certainly paradise for those on the island when our ship arrived. As soon as we dropped anchor in the lagoon, tools were downed and all available residents feasted on our company and conversation. Some came aboard to drink beer non-stop for the day and a half of our stay.

A German/Australian road constructor uses his sojourn here to the full. He shoots pig and pigeon, goes undersea fishing and sails to other islands. Studying photography in his spare time, he teaches his cook boy to prepare handsome meals of local caught and shot produce, with crayfish from the coral reef a great speciality.

We have taken on a new passenger for the last lap to Australia. His name is Mr. Filewood and is, or was, the boss of the island's logging operation.

One day he found an old anchor on the coral reef. Stories

handed down by the native population of some two to three hundred souls, related that a French sailing ship, La Perouse, having arrived in Australia shortly after Captain Cook, and been obliged to leave, sailed north-west into the South Pacific and foundered on the Vanikoro reef. No-one survived the accident, or possibly, if they did, there would have been no escape from the attentions of the hostile natives who inhabited the island at that time.

So Mr. Filewood obtained permission from both the French and British governments to blast away the coral that covered the wreck. With plenty of dynamite, supplied by those governments, he recovered cannons, anchors and coins, etc., to establish that the vessel was indeed La Perouse.

Using an old water tank and whitewashed stones, a monument was erected to commemorate the dead. Using a sharp pair of scissors, a lead plaque was suitably inscribed in both French and English. When complete, the New Hebridean Ambassador and his English equivalent, with a guard of honour made up of French sailors, attended a memorial ceremony. The gun salute was much enjoyed by the excited Vanikoran population. A feast was then prepared of crocodile steak, turtle liver and turtle curry. And Vanikoro now had a history.

When we had offloaded stores and decanted our inebriated guests ashore, we sailed, still inside the coral reef, to an almost landlocked volcanic lagoon, surrounded by wooded peaks above and crocodile-infested mangroves at sea level. Here we took aboard Kauri logs that had been floated out to our ship as rafts. Kauri trees grow in very few places and are much valued for their timber. In Vanikoro they grow in quantity.

After that, we navigated our way out to the ocean through the gap in the reef, and are now off on a short journey to Espiritu Santo, in the New Hebrides. Jim.

CHAPTER 53

Espiritu Santo, Palikoulo and Vila, New Hebrides

I had been warned that life could be a little chaotic in Espiritu Santo as the administration was shared by both the French and English. The Catholic French held sway on the day we arrived. It was Sunday, a day when they decreed that no dock work was to be done. So we were forced to spend an idle day, as were those who worked on the copra boats.

The town of corrugated-iron houses, a few shops and two bars, was not of great interest. So I headed for a large copra plantation, partly worked by natives from the nearby Banks Islands. When I was talking to one of these natives, a clatter of bird wings came from high up in a nearby bush. Two small birds had been caught in a giant spider's web, and the spider had run out of its lair to grab them. My native companion found a pole and knocked down the web, releasing the birds.

In the plantation, where trees had been planted in straight lines, ripe coconuts that had fallen from palms were gathered into piles and split in half by the blow of an axe. The flesh was then gouged out and bagged for drying and processing. This work was done by Tonkinese, originally imported from French Indo-China.

Cocoa was another product of the island. The beans were extracted from the pods where they grew and put in a bin to ferment. Having been turned over every day for a week, they were placed on a pierced metal surface and roasted by heat from below. They were then ready for export.

After our dock work was concluded on Monday, we were at sea for only half an hour before reaching Palikoulo, also in the New Hebrides. Here was a Japanese built and operated fish-freezing plant. Their vessels scoured the seas to catch tuna. They returned with the fish for hard-working natives to freeze and store them.

900 tons of the frozen fish came straight from the factory on endless belts to crash down into our freezer hold. The tuna would stay on board until the ship returned to Japan, and there be put into tins and exported - to Australia.

Still in the New Hebrides group of islands, we were to dock in Vila before heading for Brisbane. Here the natives were dressed in the more floral style of Polynesia, as mixed with the local population were Tahitians and Fijians.

Vila, being a condominium, the inhabitants were given the choice of living under French or British law. The choice seemed to be between the just laws of the British or the lesser penalties of the French. A native, charged with some petty offence, had chosen

French law, but arrived at the court half an hour late. He was told that the day's work had been concluded. Presenting himself at the court first thing the following morning, he was informed that he had already been judged and given a six-month jail sentence. He had made an unwise choice.

As we left the tropical harbour, a large turtle appeared to wave us goodbye from the clear water below our ship. The air soon became cooler. Fans were turned off and windows closed. Blankets appeared on the beds.

It was a cold and clear day on the 31st of July 1959 when pelicans escorted the Aros into Brisbane. I had reached Australia.

CHAPTER 54

Brisbane to Goondiwindi, Australia

31 July 1959

First ashore in Brisbane, I was tackled by the press for a story, which I gave in return for a lift to the hospital for a lung check. Then it was time for my first priority, to buy a good map of this part of the world, and then to find some of the fabled Australian oysters to eat.

In Brisbane I have been surprised by the cold weather (I had imagined Brisbane to be sub-tropical), by the masses of pelicans, the cleanliness of the streets and the smartly dressed people.

Aboard ship I had written the word BOURKE in large letters on a piece of cardboard. So, wasting no time, I took a tram to the outskirts of the town in the general direction of the outback, and took up position on the roadside with the said notice in one hand and a thumb pointing west from the other.

Not yet aware of Australia's vastness, I might well have been standing in Piccadilly Circus with a placard showing the name of a remote village in the north of England. So suburban cars sped by.

At last, an ancient car passed by slowly and then stopped for me. The owner, a young man, had been to Brisbane to make his fortune, failed, and was on his way back to the country. The car had two working gears — both forward. I was its reverse gear. As the brakes failed with some regularity and we crashed into other vehicles, it was my task to extricate our pile of metal from the cars we hit, and enjoy the heated arguments that the bangs engendered. After doing my job several times, I came to learn something of the expressive language used by the natives of the country.

Once on the open road, and with no other cars about, we chugged along happily, with the driver making violent corrections with the steering wheel after hitting the many potholes. After about a hundred miles of driving, my companion turned off the main track. I contributed to the cost of petrol, and stood once more on the roadside to await the arrival of another car.

Having crossed the north-south coastal mountain range, I now found myself in a monotonously flat but fertile land. There were few trees around, but many birds. And to breathe in the clean air after the stuffy atmosphere aboard ship was a great pleasure.

After a long wait, another driver brought me to Dalby, where I found a room in an hotel that lies architecturally somewhere between a Victorian pub and a galleried Georgian inn. It was cold outside, and cold inside, too. But there was a large log fire in the public bar, and a glass or two of fizzy, cold beer helped to warm me up. Beer is served in 7oz glasses at a shilling a go.

I was hungry and enquired about food. It was to be served at 6.30 sharp or not at all. I got the message and sat down promptly in a freezingly cold dining room to eat an odd kind of soup, an English style curry of something, sirloin roast, which was as close to boot leather as one could get, and a wonderful fruit salad and cream. This was all washed down with some kind of hot beef drink and rum.

If the hot drink and rum helped to warm up a body used to tropical heat, a bath sent me to bed thoroughly warmed through. It was to be my first night spent in a new and fascinating land. So I expected a good night's rest. However, in the cubicle room next to mine lived some drunks, who crashed around so much in the night that I fully expected them to fall through the wall at any time.

Breakfast was the usual outback fare of steak and eggs, with the eggs edible and the steak not. So I made enquiries about the meat. 'It's always tough here, mate. Edible stuff goes for export.' So that solved it. Jim.

2 August 1959

Dalby was about a hundred straight-line miles to the west of Brisbane. My destination of Bourke was about 500 miles to the west, but quite a bit south. So I was straying too far in a northerly direction. With this in mind I picked my next road out of town.

It was cold and foggy, but after only half an hour's wait, a very non-communicative shopkeeper picked me up. At least the five miles of our shared journey was roughly in the right direction.

Next along the dirt road came a rough-looking farmer. After nine miles he left me at an unmarked crossroads with no habitation in sight. But there was a tree nearby in which a kookaburra (a laughing jackass bird) was making plenty of noise. So I had company as he laughed away - probably at me for standing there alone in the outback, hoping to get to a place called Bourke.

Although the nights are cold, the days soon warm up and became very hot.

No-one has passed me without stopping. So, although I might have to wait all day, and possibly more, I can always be sure of a lift when the next car comes along. These waits give me time to write.

I was about to prepare for a cold night in the open when along came a brand new car in which sat a most dapper man. Known variously as King, Duke or Dick, he was a professional gambler on his way to Goodiwindi to make a killing, firstly as a bookie at the horse races in the daytime and then at the two-up card game during the night.

The 'black soil' roads on which we travelled were made of the same material as the earth in a garden. When the weather is dry and hot, the surface is graded to become flat and smooth. So the roads are then fine for speedy driving. But when rain falls, the cars sink straight in up to their axles, and the occupants either stay there or abandon their vehicles and make for the nearest hotel to wait for sunshine and drying winds. Fortunately, the weather is set fair.

We were driving through the bush when, sitting a bit

away from the roadside by some scrub, was a down-and-out-looking tramp and a youth. They were huddled round a billycan, suspended over some embers. Their clothes were dirty, and in such a poor state of repair that the seams barely held together. One could hardly imagine a more unprepossessing duo. We stopped and walked over to them.

'Hello, Charlie,' said Duke as he produced an enormous wad of notes from his pocket. 'I owe you a hundred and fifty pounds. Do you want it now or shall I put it in your bank account?' He then started to count out the money.

But Charlie decided that he didn't really want the cash, and that it would be better if it was put into his bank account. But he couldn't remember the name of his bank. So Duke agreed to keep the money until they met again.

'But have you got a drink on you, Duke?'

Duke hadn't, but I still had some aquavit left over from the sea voyage. So we sat around drinking it — from the blackest mug ever to reach my lips.

When on the road again, Duke told me that he knew the old man wouldn't remember the name of his bank, and would certainly not take the money with such a thieving scoundrel as his companion. Isolated bush country was not the place where one carried large sums of money.

I enquired about the old man. He had once been prosperous. His first financial success had been when he stole a hundred bullocks, drove them fifty miles away, and then telephoned the owner of them to say that he had located the cattle and could probably get them back. How much would it be worth? He got ten pounds a head. It was as simple as that.

I had expected the outback races at Goondiwindi to be on the rough side, but not a bit of it. Everyone was dressed in their finery and looking very smart. The occasion was clearly an important one for everyone in the district.

Duke and his fellow bookmakers each kept four books going at the same time. One was for the races at Goondiwindi, and the others for radio broadcast meetings in Sydney, Melbourne and Brisbane.

Money changed hands very quickly and in great amounts.

When I came to put a modest sum on a horse to win, the bookmaker told me, in effect, to go away and play marbles.

Always introducing myself as a Pommie, I got on famously with those around me. At the hotel where Duke and I were staying, I did my usual 'Pommie' bit. After an evening of eating and boozing, a voice boomed out from nearby, 'He's no Pommie!' But a greater compliment was to come. In the hubbub of noise, shrieking girls and drunken men (just like a Hollywood western), I met a lovely old boy called Perse Green. He kept trying to force ten pounds on me, to help 'stake me' in Australia. He didn't want me to leave. When I told him that I had decided to go home to England, he said: 'Then if you're going back, mate, don't tell them about Goodiwindi. We don't want any of them Pommies coming here to spoil it.'

Perse had been in Brussels during the First World War. In his pocket he carried his most treasured possession, a souvenir of that conflict. It was a thoroughly dog-eared postcard of the famous sculpture fountain of a small boy peeing, Manneke Pis.

Around me were a hard drinking, hard living, hard gambling and hard wenching bunch of fellows. And the ladies of pleasure looked uncommonly tough as well. Many of these men, finding themselves to be misfits in the city, or having fallen foul of the law, had successfully remade their lives in the outback.

Duke had warned me to keep clear of the action as the evening wore on. It might get a bit rough. So, with an early start for the next day in mind, and when drinking my final beer at the bar, a man sidled up to me and said: 'If anyone picks a fight with you, don't worry. Your mates'll fix 'im.' Which was as good a time as any to retire for the night.

I had not had the chance to thank Duke for his lift and kindness. So, as his room was next to mine, I tapped on his door in the early morning to say goodbye. The door was unlocked, so I peeped in, to find myself looking straight down the barrel of a loaded and cocked revolver. I had forgotten the outback rules of self-sufficiency and self-defence.

After a breakfast of steak and eggs, I was off once more, into a land where men wore hats, and swear words embellished every sentence. In the sky and in the trees around were eagles, laughing jackasses, budgerigars, galahs and others of the parrot family. It was great, and I was happy in the outback. Yet all of those to whom I have spoken, wonder why I wanted to go to Bourke, instead of the coastal towns with their bright lights, pretty girls and music. Jim.

CHAPTER 55

Goondiwindi to Bourke

4 AUGUST 1959

Locating my position on maps of Queensland and New South Wales was most confusing when towns could turn out to be one or two houses, and main roads mud trails. But to head in the general direction of Bourke was all that was necessary.

Two youths very kindly took me from the hotel to the road leading south-west. My feelings now were to hope for a long ride when waiting as near as possible to a town, just in case no car came along. Hotels were far more preferable to being outside for a cold night.

After a six hour wait, a sheep farmer not only gave me a lift for seventy five miles, but took me to see his homestead — one that was fifty miles from the nearest town. Isolated it may have been, but he had a swimming pool, deep freezer and other modern conveniences.

The landscape around the house and farm may have been desolate, but the wild life inhabitants made it a lively one. From the homestead I could see emus, wild geese, galahs, kookaburras, eagles, parakeets, and by the livestock watering lake, masses of ibis.

Outback 'towns' have the look and atmosphere seen in American cowboy films. The stars and extras here are real-life cowboys, herding cattle and sheep over flat grasslands. But the Australian cast is meaner, rougher, tougher, dirtier and smellier. Hollywood actors would rush for cover if any of these men appeared on the set. Aussie jaws and digestive systems must be tougher as well, to deal with the leathery steaks and the lashings of Worcestershire sauce, applied to anything edible, even bread and butter.

The next town in my path was the dusty, yet clean one of Moree. I was taken there by a farmer's son, who was about to play in a rugby league match nearby. The contest drew a large, enthusiastic and vociferous crowd of spectators.

Pubs were always open whenever I needed a beer. But in one hotel where I was staying on a Sunday, there was no sign of a bar, yet many people were drinking. So I followed a thirsty-looking customer to a plain door. It was opened, and money exchanged hands for beer. My turn came. 'Only travellers served on Sundays, mate.' 'But I am a traveller. I'm even a signed-in resident of this hotel.' A beer was mine. By seven in the evening most of the town's population had signed on as 'travellers'. But at eight o'clock my request for a schooner of beer was refused. 'Sorry, mate. We should have closed at six.' But I did get the beer, and at a cheaper rate than before. Perhaps it was because I was the only genuine traveller there.

Three more lifts moved me 150 miles nearer to Bourke. The first of these took me barely 30 yards down the road. The second was with two of the smelliest people I have ever met. And then, after a long wait in a desolate spot, two builders gave me a lift in their van. They were delivering materials to a school in the distant town of Goodooga. It was a hitch that would take me well out of my proposed route.

But my companions assured me that they would leave me where I might have a good chance of getting a lift to Bourke.

We approached a pond. The van was brought to a sudden halt, and out jumped the driver to take a loaded shotgun from among the bags of cement at the rear. He shot a duck. The bird fell to its death in some brackish water. Then we were off again. I enquired why he had not recovered the bird for the pot. 'Too many poisonous snakes, mate.'

I mentioned that I had not yet seen a kangaroo. And no sooner I had spoken than we were among crowds of them, lolloping around the place.

We had not travelled far, when our marksman was out of the van again, shooting left right and centre at the kangaroos — missing some, peppering and maiming others. 'Why are you shooting at them?' I asked. 'Oh, I thought you wanted to see one,' he replied.

The 'roos', although protected in some areas, do an enormous amount of damage to crops, and are shot whenever possible.

The builders left me at Goodooga, a gathering of a few houses, a pub, two stores, a school, a Saturday open-air cinema if the weather was good, and a crossroads. Near to the gate of a bungalow stood a notice saying: 'Anyone found in this garden after dark will be shot at'. But the town did have the luxury of a once-a-week air mail service. And so flat was the surrounding countryside, that an aeroplane could have landed anywhere without hazard.

I shared a room in the Goodooga pub with a miserable fellow, who thought that I might steal from him. The feeling was mutual. A certain amount of honesty was clearly expected in the outback as I had not seen a bedroom door lock since arriving in Australia.

It was a sunny and windy day when I set out again for Bourke. After a wait of three hours (the usual sort of time), a doctor collected me on his way westwards. He was to meet a friend driving toward us, and suggested that I transfer to the other car, which would retrace its steps, and continue toward Bourke via a town called Brewarrina.

Although I had seen one or two emus close-to when waiting for lifts, now, by the side of the road, was a fine flock of these large, flightless birds. 'They are liable to do anything, and it is sure to be silly,' said my doctor companion. So we slowed to a snail's pace. And they were silly. With a sudden rush, they stampeded across the road right in front of us.

We duly met the other car. After an exchange of goods and conversation, both cars turned around, and I carried on toward Brewarrina. Bourke was now within striking distance. Then I was left once more at the roadside.

After a couple of hours, the boss of a sheep shearing company, called John Barwood, and his wife, not only took me through the town of Brewarrina but on to give me a glance of my target town, Bourke. They then drove me back to the hotel in Brewarrina, from where they offered to take me the following day to draw men shearing sheep.

At six in the morning I was collected from the hotel. The land all around was, as had been most of the country so far, flat and almost treeless as far as the eye could see. Only emus and hopping or sitting kangaroos punctuated the landscape of red sand and dry Mitchell grass. This grass is considered to be the finest in the world for fattening and rearing sheep.

After much driving, we reached the raised shearing sheds. The sheep were driven up from below and grabbed by the shearers, who held the beasts down and swept over them with clattering, venomous-looking, mechanical shears. Cut and bleeding from surface wounds, the skinny-looking animals were then shoved down a ramp into pens at the rear. Their worries were not over. When released into the wilds once more, they could be the prey of dingoes, eagles or carnivorous wild pigs.

P.R

Many of the best shearers were aborigines. In fact, much of the population hereabouts varied in colour from black to almost white. A white shearer told me: 'The longer you are out here, mate, the whiter the girls seem to become.'

Wages are good, but with petrol costs high, and cars in constant use on bad roads, and thus in regular need of repair, the cost of living is high, too. And telephone calls are all long distance.

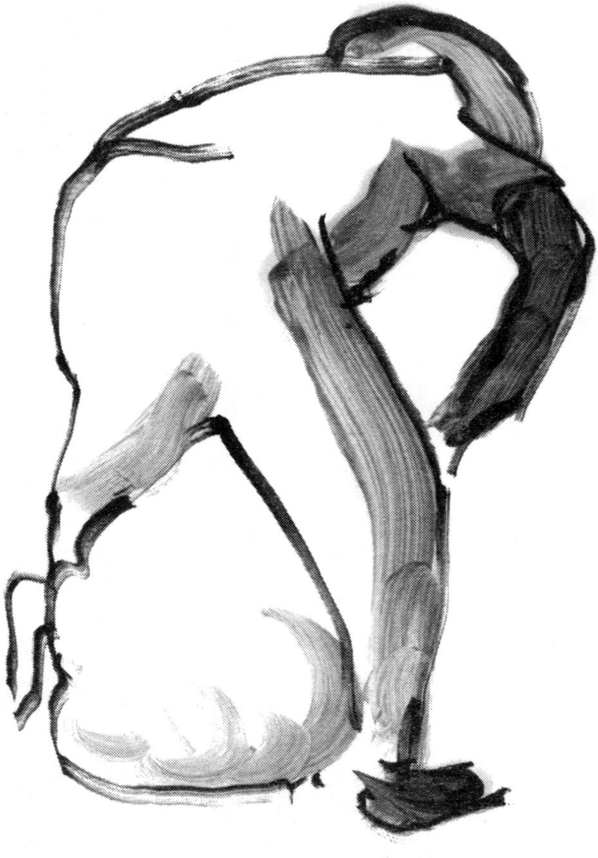

We visited a homestead that had been completely cut off for three months when rains turned roads into a quagmire. But as soon as the weather became dry, out came the graders to level and smooth the road surfaces once more.

John, a typical go-ahead and imaginative outback Aussie, had bought himself some land by the Darling River. On it he was building a house and growing tobacco and vegetables.

As the natural rainfall was quite inadequate for nourishing anything other than Mitchell grass, he used an old tractor to pump water from the river for irrigation. With an excellent climate and adequate water supply, almost anything could be grown. In fact, the area had once been famous for its oranges. And although the locally grown orange I tasted was quite delicious, only one orchard remained in the district. The cost of transportation to distant markets made growing them uneconomical.

It is around the 6th or 7th of August 1959. I have returned to Bourke, and find the town to have style and architectural charm. It was once a prosperous place and used as an inland port for shipping wool down the Darling River. Now, with the river no longer navigable, it has the atmosphere of a ghost town.

A Port Authority Customs House, and wrecks of riverboats, and even drawbridges, still exist as indications of its past. It is a place where I would dearly like to stay. But with my goal attained, I find I have barely enough money to get back to England. Wisdom dictates a speedy return via America.

I set out early tomorrow for Sydney. Although I am only half way around the world I have reached the last lap. Jim.

CHAPTER 56

Bourke to Sydney, New Zealand, Tahiti, Panama Canal, Florida, New York and London

Hitch-hiking to Sydney (arrival 9th August 1959) was notable for the hospitality of an Italian wine merchant in cold Orange. He not only invited me into his shop and home to warm up after standing in icy wind and rain, but also offered me a taste of some Australian wines. Since arriving in the country, and travelling in the outback, wine had been a drink taken only by women and sissies. Beer was the man's drink. I now had a chance to sample the liquid in which I take so much pleasure. Penfolds Grange Hermitage '53, bottled in '54, was an absolute revelation. Nothing approaching it had ever entered England to my knowledge. I was to learn later that the better Australian wines were bought by those in the know before they had even been made. Good wine never left the country.

Sydney I found to be a fine and vibrant city, and with unbelievably good fish and shellfish on offer in restaurants and the houses of friends. Fresh orange juice was sold by the pint. The pictures in the Sydney Art Gallery were of a high class, with fine, early Sydney Nolans and a wonderful landscape by Russell Drysdale, who had another fine one in the Terry Clune Gallery. 'Tas' Drysdale and I met, and got on famously. We would both have gone to the Northern Territories if I had had the money and time. A notice in a lift said: 'Gentlemen are requested to keep their hats on in the lift'. The foundations had just been laid for an opera house by the harbour. There were lotteries to help pay for it.

I had travelled 1,600 miles through the outback and loved every bit of it. Now it was the 20th of August, 1959, and time to board the old and ill-fated Dutch liner, Johan van Oldenbarneveldt (she sank later, under another name, off the Canary Islands).

Our first port of call was Aukland, in New Zealand. The soft and lovely countryside of undulating, sheep-covered hills, the cleanliness, tidiness, and the lack of dirt, vagrants or vice, came as a contrasting shock after the Far East and Australia. But it was clearly a paradise country for those appreciating orderliness and the great outdoors.

Papeete, in Tahiti, could hardly have changed since Gauguin's day. There were cafés for drinking and dancing, with garlanded Tahitians jumping around like mad. A girl on board danced with one of them and nearly had her ear bitten off. It was a lovely island, full of life, gaiety and pretty women. Girls still bathed naked in mountain streams. I walked alone around the coast and tropical hills, and bought one of my few souvenirs — a flat shell, from Gauguin's son.

The Panama Canal was hot and steamy, with its huge locks swallowing up our large ship. Little railway engines pulled us through the open lock gates to a series of inland lakes. In Cristobal, at the far end, a curvaceous negress stood beside a murky doorway and said: 'You want me? I give you drink, and cigar, too.' The Dutch food aboard was mostly covered with nutmeg, and seemed even less palatable in the Central American heat. A madman smoked lavatory paper in his pipe. But then it was a very mixed crowd. I suppose it was bound to be with 1,210 passengers aboard.

Disembarking in Florida, after a tediously long 'cruise' voyage, I took a Greyhound Bus journey to New York.

Although jets had just been introduced for the Atlantic crossing, at extra cost, I was only able to take a turbo-prop Britannia back to London.

October 1959.

With £1000 well spent, I had furthered my education, circumnavigated the world, seen more, experienced more, drawn more (1500 drawings, with many destroyed by me later as being not good enough), and done more than I had ever thought possible. I had met wonderful and hospitable people (and some unpleasant ones), and been befriended at every turn. I had also felt lonely, been worried, and frightened many times. Acquiring visas, and extensions to them, was never easy or pleasant. Some parts of the world, and many of the people that I found to be of interest, I was able to record in the drawings. The 110 letters of details, sent back to England, formed the basis of this book. I hope you have shared in the pleasure that I have taken in writing it.

SOUTH-EAST ASIA

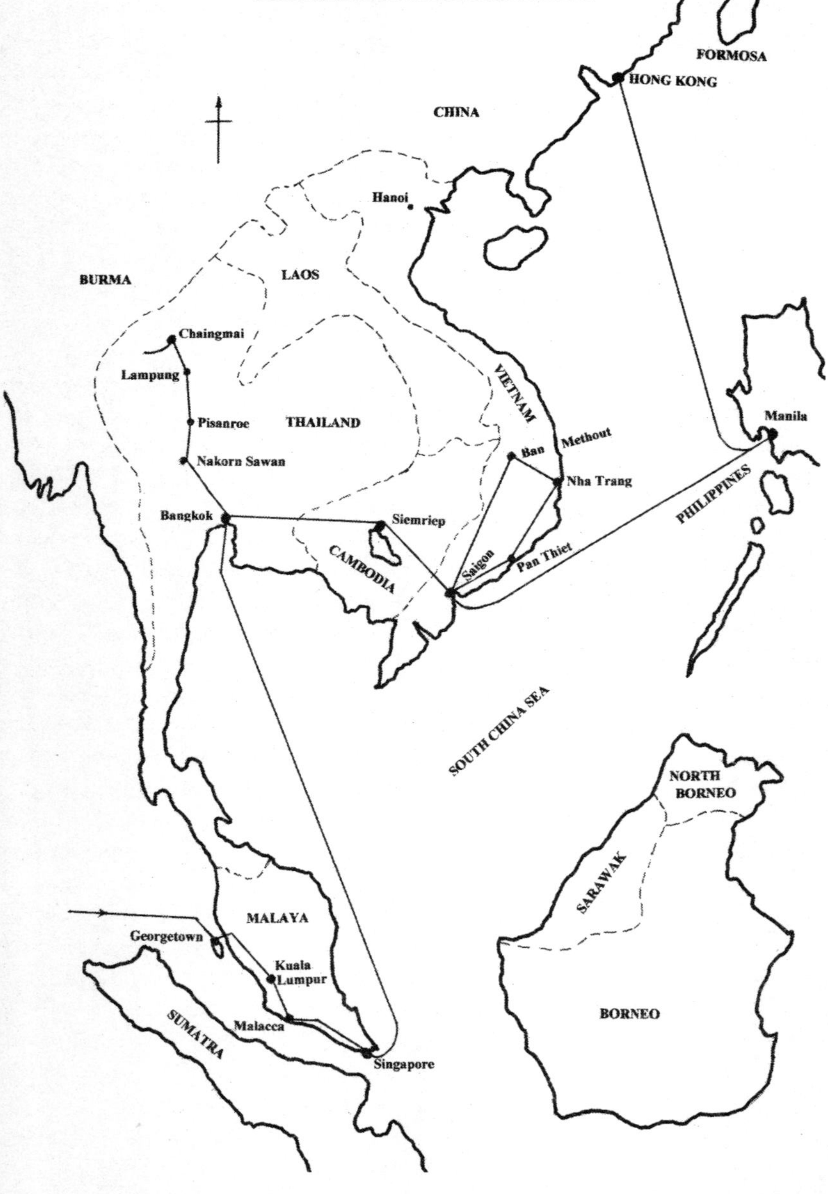

JAPAN

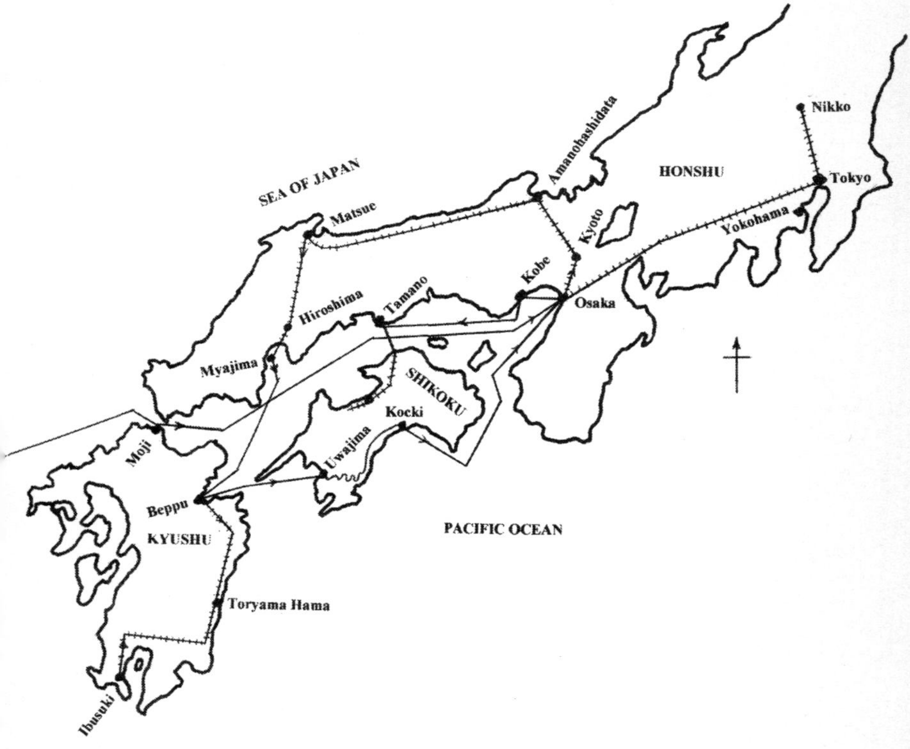

SEA OF JAPAN
Matsue
Amanohashidata
HONSHU
Nikko
Tokyo
Yokohama
Kyoto
Kobe
Osaka
Hiroshima
Tamano
Myajima
SHIKOKU
Kocki
Uwajima
Moji
Beppu
KYUSHU
Toryama Hama
Ibusuki
PACIFIC OCEAN

THE SOUTH PACIFIC

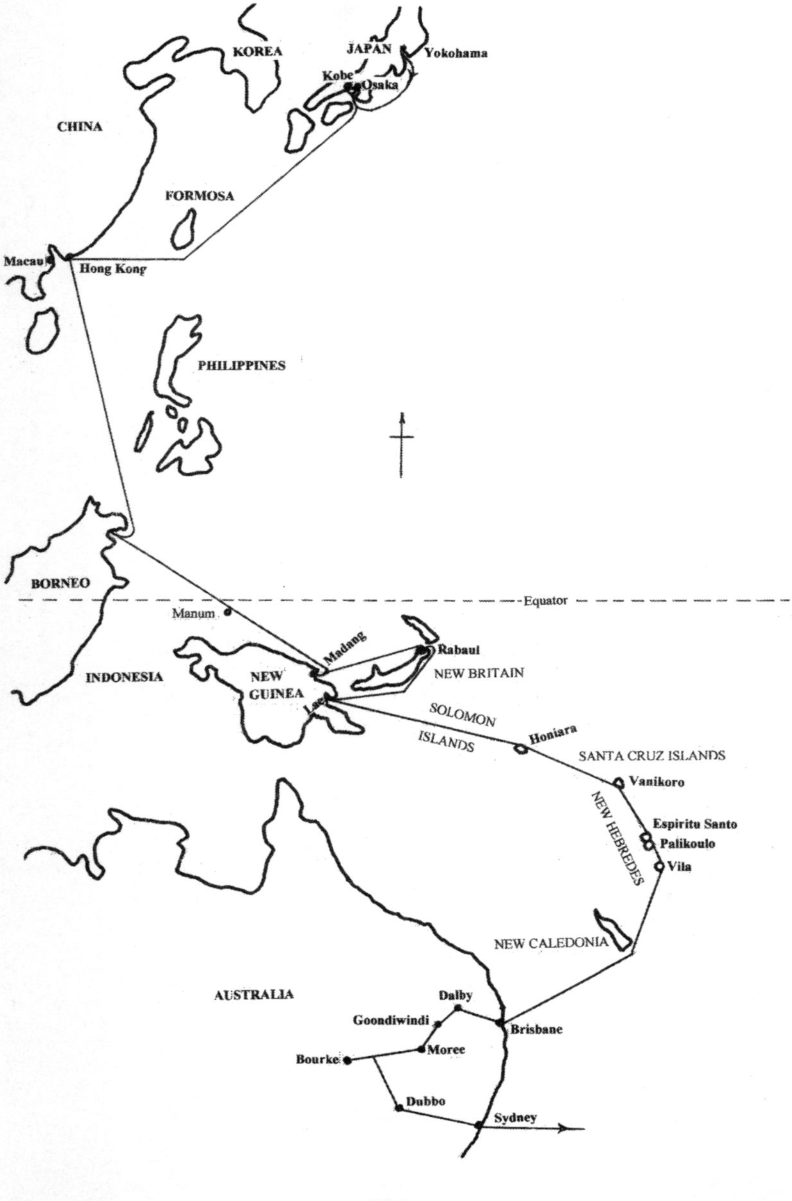

TRAVELLER'S NOTES AND DRAWINGS

TRAVELLER'S NOTES AND DRAWINGS

TRAVELLER'S NOTES AND DRAWINGS

TRAVELLER'S NOTES AND DRAWINGS

TRAVELLER'S NOTES AND DRAWINGS

TRAVELLER'S NOTES AND DRAWINGS

TRAVELLER'S NOTES AND DRAWINGS

TRAVELLER'S NOTES AND DRAWINGS

TRAVELLER'S NOTES AND DRAWINGS

TRAVELLER'S NOTES AND DRAWINGS

TRAVELLER'S NOTES AND DRAWINGS

TRAVELLER'S NOTES AND DRAWINGS

TRAVELLER'S NOTES AND DRAWINGS

TRAVELLER'S NOTES AND DRAWINGS

TRAVELLER'S NOTES AND DRAWINGS

TRAVELLER'S NOTES AND DRAWINGS

TRAVELLER'S NOTES AND DRAWINGS

TRAVELLER'S NOTES AND DRAWINGS

TRAVELLER'S NOTES AND DRAWINGS

TRAVELLER'S NOTES AND DRAWINGS

TRAVELLER'S NOTES AND DRAWINGS

TRAVELLER'S NOTES AND DRAWINGS

TRAVELLER'S NOTES AND DRAWINGS

TRAVELLER'S NOTES AND DRAWINGS

TRAVELLER'S NOTES AND DRAWINGS

TRAVELLER'S NOTES AND DRAWINGS

TRAVELLER'S NOTES AND DRAWINGS

TRAVELLER'S NOTES AND DRAWINGS

TRAVELLER'S NOTES AND DRAWINGS

TRAVELLER'S NOTES AND DRAWINGS

TRAVELLER'S NOTES AND DRAWINGS

TRAVELLER'S NOTES AND DRAWINGS

TRAVELLER'S NOTES AND DRAWINGS